Wastewater Treatment: Concepts and Practices

Wastewater Treatment: Concepts and Practices

Ronald Hogan

SYRAWOOD
PUBLISHING HOUSE

New York

Published by Syrawood Publishing House,
750 Third Avenue, 9th Floor,
New York, NY 10017, USA
www.syrawoodpublishinghouse.com

Wastewater Treatment: Concepts and Practices
Ronald Hogan

International Standard Book Number: 978-1-68286-825-6 (Hardback)

Cataloging-in-Publication Data

Wastewater treatment : concepts and practices / Ronald Hogan.
    p. cm.
Includes bibliographical references and index.
ISBN 978-1-68286-825-6
1. Sewage--Purification. 2. Sewage--Purification--Environmental aspects. 3. Sewage disposal plants.
I. Hogan, Ronald.
TD430 .W37 2019
628.162--dc23

TABLE OF CONTENTS

xx

Permissions

Index

PREFACE

Wastewater treatment is a process through which wastewater is converted into usable form with minimal impact on the environment. The by-products generated from wastewater treatment plants like grit, sewage sludge and screenings, are also treated in the wastewater treatment plant. It involves various phase separation processes such as sedimentation and filtration, polishing, biochemical oxidation and chemical oxidation. Treatment of wastewater depends on the type of wastewater. Accordingly, wastewater treatment plants can be categorized as industrial wastewater, agricultural wastewater, sewage and leachate treatment plants. This book outlines the processes and applications of wastewater treatment in detail. It presents this complex subject in the most easy to understand language. This book attempts to assist those with a goal of delving into the field of wastewater treatment.

Given below is the chapter wise description of the book:

Chapter 1- The water, which is a byproduct of any industrial, commercial, agricultural or domestic activities, is termed as wastewater. This is an introductory chapter, which discusses in detail the varied types of wastewater, especially sewage and industrial wastewater. It further elucidates their sources in order to provide a comprehensive understanding.

Chapter 2- Wastewater quality indicators are the tests and techniques implemented for assessing the suitability of wastewater for re-use or disposal. These tests measure biological, chemical and physical characteristics of wastewater. This chapter has been carefully written to provide an easy understanding of the varied characteristics that are measured by such quality indicators, like biochemical oxygen demand, chemical oxygen demand, total dissolved solids, total suspended solids, etc.

Chapter 3- The treatment of wastewater is an important aspect of sanitation. It includes the management of human and solid waste, industrial wastewater and sewage treatment, stormwater management, etc. The aim of this chapter is to explore the different techniques of treatment of wastewater such as coagulation water treatment, secondary and tertiary wastewater treatment, use of API oil-water separator, septic tank, etc.

Chapter 4- Science and technology has undergone rapid development in the past decade, which has resulted in the emergence of many innovative tools and techniques for wastewater management. These include effective wastewater management systems such as onsite sewage facility, aerated lagoon, decentralized wastewater system, grinder pump and grease trap, which have been extensively detailed in this chapter.

At the end, I would like to thank all those who dedicated their time and efforts for the successful completion of this book. I also wish to convey my gratitude towards my friends and family who supported me at every step.

Ronald Hogan

Chapter 1

Wastewater: Types and Sources

The water, which is a byproduct of any industrial, commercial, agricultural or domestic activities, is termed as wastewater. This is an introductory chapter, which discusses in detail about the varied types of wastewater, especially sewage and industrial wastewater. It further elucidates their sources in order to provide a comprehensive understanding.

Wastewater refers to all effluent from household, commercial establishments and institutions, hospitals, industries and so on. It also includes stormwater and urban runoff, agricultural, horticultural and aquaculture effluent.

Effluent refers to the sewage or liquid waste that is discharged into water bodies either from direct sources or from treatment plants. Influent refers to water, wastewater, or other liquid flowing into a reservoir, basin or treatment plant.

Wastewater from farming tends to be dealt with separately. Increasingly, farmers are being required to set up on-site treatment systems for such things as animal effluent. Households, industries and commercial businesses can all use on-site systems, but oen, depending on the size of the community, their wastewater will be combined and managed together.

This means that the wastewater in your area will be unique. An essential factor in determining the kinds of wastewater you will need to deal with will be the kinds of industries and processing businesses in your area. For example, if there is a local cheese factory, your wastewater system will have to deal with whey as a waste. If there is a metal-processing factory, your system may have to deal with water that has been used to wash down machinery.

Wastes from industry and businesses are known as tradewastes. It will be important to take account of these tradewastes when designing your system, and important to take account of initiatives being undertaken by industry to reduce the volume and toxicity of their wastes. It would be worth working with these industries in order to help them to deal with their own waste streams.

Sewage is also wastewater. It is wastewater originating from toilets and bathroom fixtures, bathing, laundry, kitchen sinks, cleaners, and similar dirty water that is produced in households and public places. Water used to irrigate turf and gardens, swimming pools, roof drainage, surface runoff and stormwater are all wastewater but not classified as sewage.

In simple terms, wastewater is all the dirty water from municipal sources (poop, urine and faecal sludge). This includes black water, gray water and yellow water. All dirty water from all the schools, restaurants, commercial establishments, hospitals, farms, floodwater and all the possible dirty water you can think of is considered wastewater. Some wastewater contain hazardous dissolved toxins and chemicals, whiles others contain particles, sediments and suspended matter of all sizes.

Agriculture (irrigation, livestock watering and cleaning, aquaculture) uses about 69% to 90% of global fresh water use, and the bulk of it is returned to the soil, waterways or discharged with added nutrients and contaminants.

Sources

Sources of wastewater include the following domestic or household activities:

- Human excreta (feces and urine) often mixed with used toilet paper or wipes; this is known as blackwater if it is collected with flush toilets;

- Washing water (personal, clothes, floors, dishes, cars, etc.), also known as greywater or sullage;

- Surplus manufactured liquids from domestic sources (drinks, cooking oil, pesticides, lubricating oil, paint, cleaning liquids, etc.).

Activities producing industrial wastewater:

- Industrial site drainage (silt, sand, alkali, oil, chemical residues);

- Industrial cooling waters (biocides, heat, slimes, silt);

- Industrial processing waters;

- Organic or biodegradable waste, including waste from hospitals, abattoirs, creameries, and food factories;

- Organic or non bio-degradable waste that is difficult-to-treat from pharmaceutical or pesticide manufacturing;

- Extreme pH waste from acid and alkali manufacturing;

- Toxic waste from metal plating, cyanide production, pesticide manufacturing, etc.;

- Solids and emulsions from paper mills, factories producing lubricants or hydraulic oils, foodstuffs, etc.;

- Water used in hydraulic fracturing;
- Produced water from oil & natural gas production.

Other activities or events:

- Urban runoff from highways, roads, carparks, roofs, sidewalks/pavements (contains oils, animal feces, litter, gasoline/petrol, diesel or rubber residues from tires, soapscum, metals from vehicle exhausts, de-icing agents, herbicides and pesticides from gardens,etc.);
- Agricultural pollution, direct and diffuse.

Wastewater can be diluted or mixed with other types of water by the following mechanisms:

- Seawater ingress (high volumes of salt and microbes);
- Direct ingress of river water;
- Rainfall collected on roofs, yards, hard-standings, etc. (generally clean with traces of oils and fuel);
- Groundwater infiltrated into sewage;
- Mixing with other types of wastewater or fecal sludge.

Pollutants

The composition of wastewater varies widely. This is a partial list of pollutants that may be contained in wastewater:

Chemical or Physical Pollutants

- Heavy metals, including mercury, lead, and chromium;
- Organic particles such as feces, hairs, food, vomit, paper fibers, plant material, humus, etc.;
- Soluble organic material such as urea, fruit sugars, soluble proteins, drugs, pharmaceuticals, etc.;
- Inorganic particles such as sand, grit, metal particles, ceramics, etc.;
- Soluble inorganic material such as ammonia, road-salt, sea-salt, cyanide, hydrogen sulfide, thiocyanates, thiosulfates, etc.;
- Macro-solids such as sanitary napkins, nappies/diapers, condoms, needles, children's toys, dead animals or plants, etc.;
- Gases such as hydrogen sulfide, carbon dioxide, methane, etc.;
- Emulsions such as paints, adhesives, mayonnaise, hair colorants, emulsified oils, etc.;
- Toxins such as pesticides, poisons, herbicides, etc.;
- Pharmaceuticals and hormones and other hazardous substances;

- Thermal pollution from power stations and industrial manufacturers.

Biological Pollutants

If the wastewater contains human feces, as is the case for sewage, then it may also contain pathogens of one of the four types:

- Bacteria (for example *Salmonella*, *Shigella*, *Campylobacter*, *Vibrio cholerae*),

- Viruses (for example hepatitis A, rotavirus, enteroviruses),

- Protozoa (for example *Entamoeba histolytica*, *Giardia lamblia*, *Cryptosporidium parvum*) and

- Parasites such as helminths and their eggs (e.g. *Ascaris* (roundworm), *Ancylostoma* (hookworm), *Trichuris* (whipworm)).

It can also contain non-pathogenic bacteria and animals such as insects, arthropods, small fish.

Quality Indicators

Since all natural waterways contain bacteria and nutrients, almost any waste compounds introduced into such waterways will initiate biochemical reactions (such as shown above). Those biochemical reactions create what is measured in the laboratory as the biochemical oxygen demand (BOD). Such chemicals are also liable to be broken down using strong oxidizing agents and these chemical reactions create what is measured in the laboratory as the chemical oxygen demand (COD). Both the BOD and COD tests are a measure of the relative oxygen-depletion effect of a waste contaminant. Both have been widely adopted as a measure of pollution effect. The BOD test measures the oxygen demand of biodegradable pollutants whereas the COD test measures the oxygen demand of oxidizable pollutants.

Any oxidizable material present in an aerobic natural waterway or in an industrial wastewater will be oxidized both by biochemical (bacterial) or chemical processes. The result is that the oxygen content of the water will be decreased.

Treatment

At a global level, around 80% of wastewater produced is discharged into the environment untreated, causing widespread water pollution.

There are numerous processes that can be used to clean up wastewaters depending on the type and extent of contamination. Wastewater can be treated in wastewater treatment plants which include physical, chemical and biological treatment processes. Municipal wastewater is treated in sewage treatment plants (which may also be referred to as wastewater treatment plants). Agricultural wastewater may be treated in agricultural wastewater treatment processes, whereas industrial wastewater is treated in industrial wastewater treatment processes.

For municipal wastewater the use of septic tanks and other On-Site Sewage Facilities (OSSF) is widespread in some rural areas, for example serving up to 20 percent of the homes in the U.S.

One type of aerobic treatment system is the activated sludge process, based on the maintenance and recirculation of a complex biomass composed of micro-organisms able to absorb and adsorb the organic matter carried in the wastewater. Anaerobic wastewater treatment processes (UASB, EGSB) are also widely applied in the treatment of industrial wastewaters and biological sludge. Some wastewater may be highly treated and reused as reclaimed water. Constructed wetlands are also being used.

Disposal

Industrial wastewater effluent with neutralized pH from tailing runoff in Peru.

In some urban areas, municipal wastewater is carried separately in sanitary sewers and runoff from streets is carried in storm drains. Access to either of these systems is typically through a manhole.

During high precipitation periods a combined sewer system may experience a combined sewer overflow event, which forces untreated sewage to flow directly to receiving waters. This can pose a serious threat to public health and the surrounding environment.

Sewage may drain directly into major watersheds with minimal or no treatment but this usually has serious impacts on the quality of an environment and on the health of people. Pathogens can cause a variety of illnesses. Some chemicals pose risks even at very low concentrations and can remain a threat for long periods of time because of bioaccumulation in animal or human tissue.

Wastewater from industrial activities may be pumped underground through an injection well. Wastewater injection has been linked to Induced seismicity.

Effects of Wastewater

Organic Material

The organic content of wastewater is made up of human faeces, protein, fat, vegetable and sugar material from food preparation, and soaps from cleaning. Some of this is dissolved into the water and some exists as separate particles.

Ecosystem Health Effects

Naturally occurring soil and water bacteria eat this organic waste and use it to grow rapidly. In a natural or dilute water environment where there is plenty of oxygen dissolved in the water, aerobic (oxygen-using) bacteria eat the organic material and form a slime of new bacterial cells and dissolved salt-waste products. If undiluted wastewater is left on its own, however, anaerobic (non-oxygen-using) bacteria decompose the waste organic material and release odorous gases such as hydrogen sulphide, as well as 'non-smelly' gases such as methane and carbon dioxide.

It is the amount of oxygen removed or the too-rapid growth of the bacterial slime that can cause the harm.

The important thing is to measure how much oxygen will be used by aerobic bacteria to convert the organic material to new bacteria. This is the 'biochemical oxygen demand' (BOD), and the standard measure is the amount of dissolved oxygen needed by aerobic bacteria over a five-day period at a water temperature of 20° Celsius (called the BOD_5). The BOD_5 strength of wastewater indicates its potential polluting impact if it is not treated. It is measured in parts per million (ppm), or in the metric system the number of grams of organic material per cubic metre (g/m^3). The BOD_5 of untreated wastewater is around 200–300g/m^3, while the BOD_5 for a healthy aquatic ecosystem would be less than 5g/m^3.

Relating these scientific measurements to everyday experience, a central issue is how much oxygen is left for fish to breathe after aerobic bacteria have used the oxygen to break down the organic material. If BOD_5 levels of less than 4 g/m^3 occur in a stream that has naturally healthy levels of dissolved oxygen, then the stream system can deal with the amount of waste without affecting the fish. A good-quality healthy level of dissolved oxygen in water is around 8 to 10 g/m^3. At a dissolved oxygen level of 5 g/m^3 the fish become stressed, and at 2 g/m^3 the fish will die from lack of oxygen unless they are able to move to more oxygenated waters.

Figure: The effects of organic material and nutrients released into waterways

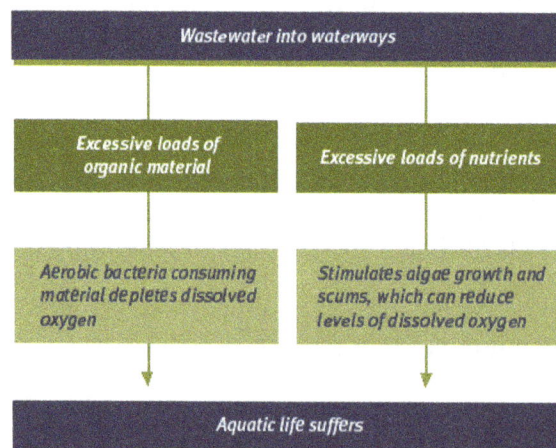

Textual Description

Where there is an overwhelming amount of wastewater, all the oxygen will be used up and the anaerobic bacteria will take over. The water will go septic (anaerobic) and the fish will die, as will other forms of oxygen-dependent life. This is partly why wastewater is treated to remove as much organic material as possible. But the content of even treated wastewater can be an issue for your community. Sensitive streams and estuaries are particularly vulnerable.

In effect, ecosystem services can be damaged, and these problems may be felt well before the level of pollution directly affects human health. For your area, you will need to know how much wastewater is entering – or may enter – your local stream or river, and the level of dissolved oxygen. Talking with the regional council may help with this.

Suspended Solids

The portion of organic material that does not dissolve but remains suspended in the water is known as suspended solids. The level of suspended solids in untreated wastewater is around 200 g/m^3.

Ecosystem Health Effects

If effluent is discharged into streams untreated, any solids it contains will tend to settle in quiet spots. Oxygen levels will soon be depleted in the area of the contamination, causing it to decompose anaerobically. If there are high concentrations of this contamination the water in the stream will go septic because the oxygen will be used up. This will not only smother the fish, but will also kill off the life at the bottom of the stream, creating dead zones.

Dissolved Salts

The most significant salts in wastewater are nitrates and phosphates. These occur naturally to some extent. Nitrate also derives from the breakdown of organic nitrogen in protein waste matter, and the oxidation of the ammonia in urine. Phosphates are present in detergents used in washing and laundering, and are also produced by organic breakdown. The total nitrate in wastewater is around 40 g/m^3, and phosphate is around 15 g/m^3.

Ecosystem Health Effects

Nitrates and phosphates are essential elements for growth. When nitrates and phosphates are discharged into natural waters they fertilise the growth of microscopic algae and water 'weeds', which can lead to green algal suspensions and weed mats. This overgrowth results in their death and decay, and means further consumption of dissolved oxygen and smothering of aquatic life. The nutrients that caused the initial growth can then be released back into the water, initiating another cycle of weed and algal growth and decay.

Bacteria and Viruses

The human gut produces a huge quantity of bacteria, which are excreted as part of faeces on a daily basis. The most common and easily measured organism is *E.coli* (Escherichia coliform group), which is referred to by wastewater scientists and engineers as 'faecal coliform' bacteria. This is called an 'indicator' because its presence indicates the presence of faecal matter from warm-blooded animals. More extensive testing is required to tell if the source is human or not.

Special tests are needed to distinguish between the amount of pollution produced by humans and the amount produced by birds and other animals that gets into the water.

The amount of faecal coliform is measured per 100 ml of water – around half a cup. Each person excretes about 140 billion faecal coliforms a day. In untreated wastewater the faecal coliforms can be around 10 to 100 million per 100 ml. It is the presence of these faecal coliforms that the drinking-water standards and recreation standards are concerned with.

The main class of viruses are the enteric viruses, which cause gastro-enteritis; for example, calcivirus (Norwalk virus), rotavirus, enterovirus (polio and meningitis) and hepatitis. Generally viruses do not replicate in the outside world, but they may survive for a long time. Spray irrigation may shock viruses into die-off due to exposure to ultraviolet light or drying out of their surroundings. Poliovirus 3 has been found in aerosols at a wastewater treatment plant .[3] In a marine environment some viruses have been known to survive a number of days, possibly protected in suspended solids.

Human Health Effects

Many of the faecal coliform bacteria in human waste are harmless. However, there are disease organisms – or 'pathogens' – that can cause harm. These can be bacteria such as typhoid, or viruses such as hepatitis B. Direct contact with these pathogens or pollution of the water supply can cause infections. The Ministry of Health has national responsibility for developing drinking-water standards, which will guide your community's understanding of the risks it might face from local wastewater. Sewage can pollute shellfish-gathering areas and, if eaten, the shellfish will cause illness. Shellfish filter food by passing several litres of water an hour through their system. The food concentrates in the shellfish, which means that any pathogens will also accumulate.

Relatively high concentrations can also make an area unsafe for swimming and 'water contact recreation'. National guidelines developed by the Ministry for the Environment help local communities to classify their harbours, streams and lakes in terms of safety for swimming, fishing and shellfish gathering. Local regional councils will set standards for discharges for these areas. These standards relate to the amount of bacteria present in a certain volume of water.

Ground water can also become contaminated. Wastes can percolate through the soils into underground water or aquifers. Given that many smaller communities and farms obtain their water from bores or wells into these aquifers, this contamination can be a serious issue.

During the nineteenth century the large quantities of sewage in the bigger towns and cities were identified as a health problem. Finding solutions to cholera epidemics from infected water supplies was a major issue. The wastewater system you now have may well be a direct heritage of these concerns.

Other Dissolved Constituents

Wastewater contains metals, chemicals and hormones from households (via food, medicines, cosmetics and cleaning products) and business processes (eg, mercury from dentistry, which can easily be removed by installing a centrifuge in dental surgeries). It can also contain halogenated hydrocarbons and aromatics, plasticisers, polyaromatic and petroleum hydrocarbons, organo-chlorine pesticides, PCBs and dioxins.

There are two issues: if large quantities are discharged into small, highly localised areas, such as a stream or small lake, there may be pollution problems. The other issue is the 'bio-accumulation' of these substances in various parts of the food chain. This can bring unacceptable concentrations in humans and aquatic life, which can lead to health problems.

Human Health Effects

Long-term Health Impacts of Residues in Water Supplies and Food

The issue here is one of long-term impacts of various wastewater residues on the human system. Water naturally contains such things as iron, zinc and manganese, but industrial processes can introduce higher concentrations. If the concentrations are high enough, exposure to some metals and chemicals may have an impact on how the body's system works.

The long-term impacts of these substances on human health are not always well understood. Wastewater will carry a range of substances, which can pass into the water supply or be returned to the soil in heavy concentrations. Some treatment systems will remove metals and chemicals from the wastewater, but the sludge produced as a result of this treatment will then contain a high concentration of these substances. The New Zealand Waste Strategy calls for such wastes, by 2007, to be beneficially used or appropriately treated to minimise the production of methane and leachate. Whatever use the sludge is put to, it should comply with the Biosolids Guidelines.

Endocrine Disruption

The endocrine system in the human body is a complex network of glands and hormones that regulate many of the body's functions, including growth, development and maturation, as well as the way various organs operate. The endocrine glands – including the pituitary, thyroid, adrenal, thymus, pancreas, ovaries and testes – release carefully measured amounts of hormones into the bloodstream, which act as natural chemical messengers. They travel to different parts of the body to control and adjust many life functions.

An endocrine disruptor is a synthetic chemical, which, when absorbed into the body, either mimics or blocks hormones and disrupts the body's normal functions. This disruption can happen through altering normal hormone levels, halting or stimulating the production of hormones, or changing the way hormones travel through the body. This is a new area of scientific investigation and is not yet well understood. There are concerns that, for example, the decline in fertility levels in all animals in the food chain, including humans, could be as a result of excessive discharge of these chemicals. Such investigations are now being considered in New Zealand.

The issue is relevant to wastewater issues because many of these substances will enter the food chain – either on land or in waterways – from wastewater. Of course some of the chemicals (eg, some pesticides) will also enter the ecosystem via run-off from farms and roadways. Wastewater treatment systems will remove some of these chemicals, but generally treatment processes are not currently designed to deal with this problem.

Ecosystem Health Effects

Endocrine Disruption

The issue raised for human health is also relevant to aquatic ecosystems. There is some concern that the hormone-producing systems in fish are under pressure. High levels of oestrogen released from wastewater can affect the reproductive cycles of fish. The degree to which this is an issue in New Zealand is not known.

Toxic Effects on Freshwater and Marine Life

These can have the immediate effect of killing fish, invertebrates and even plant life. This can be a serious loss in itself, but there are also flow-on effects. The dead fish or plants will be broken down, and can contribute to further depletion of oxygen in the water.

The key point to remember is that wastewater management is not just about toilet flushing, bathing, cooking and washing water. It is likely your community will have tradewastes, even if just from the local garage. Your overall catchment will have a huge variety of different wastewaters that will need to be considered. *Table* summarises the different components of wastewater that cause problems.

Table: The Problem-causing Components of Wastewater

Type of Material In Wastewater	Comment
Organic waste: • human waste • food waste • industrial and commercial wastes • animal effluent	faeces, urine, blood • an increasing volume of wastewater - possibly due to the advent of kitchen waste macerators • tradewaste - processing • farming - not usually managed via community infrastructure
Oils and fats	• households – usually from tipping down drains • tradewastes – garages, manufacturing
Metals	• households – found in foods – via human wastes • aggressive water supplies (outside the limits pH6-8) • tradewastes
Solvents	• households – tipping down drains, cleaning • tradewastes – garages, manufacturing
Chemicals	• households – via human wastes • households – via cleaners, soaps etc, washing, bathing and cooking • tradewastes
Paints	• households • tradewastes

Blackwater (Waste)

Blackwater is water that has been mixed with waste from the toilet. Because of the potential for contamination by pathogens and grease, water from kitchens and dishwashers should be excluded from greywater and considered as blackwater.

Each wastewater type must be treated differently and can be used in various ways. Greywater is ideal for garden watering, with the appropriate precautions, such as using low or no sodium and phosphorus products and applying the water below the surface. Appropriately treated greywater can also be reused indoors for toilet flushing and clothes washing, both significant water consumers.

Blackwater requires biological or chemical treatment and disinfection before reuse. For single dwellings, treated and disinfected blackwater can be used only outdoors, and often only for subsurface irrigation.

Properties of Blackwater

Blackwater, unlike other wastewater streams, consists of a large proportion of organic matter, and, due to high phosphate and nitrogen concentrations, is particularly rich in nutrients. However, studies show that the concentration of pathogens and micro-pollutants such as drug residues is higher in blackwater than that of other wastewater streams. A return of the water to the countryside - especially when mixed with other wastewater streams, such as greywater and rainwater - therefore requires an energy-intensive cleaning process.

But it could be done differently: A separation of the blackwater from the other wastewater streams allows for the use of alternative treatment and recovery processes. The chemical energy bound in the organic material can be utilized through anaerobic treatment. For example, the blackwater can be digested in a biogas plant, producing a source of renewable energy. Thus, the recycling of blackwater performs an ecological service and makes a contribution to the field of renewable energy.

Treatment Processes

Blackwater contains pathogens that must decompose before they can be released safely into the environment. It is difficult to process blackwater if it contains a large quantity of excess water, or if it must be processed quickly, because of the high concentrations of organic material.

Composting

However, if blackwater does not contain excess water, or if it receives primary treatment to de-water, then it is easily processed through composting. The heat produced by naturally occurring thermophilic microorganisms will heat the compost to over 60 °C (140 °F), and destroy potential pathogens.

Blackwater generation can be avoided by making use of composting toilets and vermifilter toilets. In certain autonomous buildings, such as earthships, this is almost always present and allows the water requirements of the building (which, with earthships, are self-generated) to be heavily reduced. Besides saving water, composting toilets allow the user to reuse the nutrients found therein (e.g., for growing crops/trees).

Blackwater (Coal)

Blackwater is the pollution produced from coal preparation.

It contains copious amounts of hazardous substances, including carcinogenic compounds and heavy metals. Blackwater is permanently impounded in toxic waste facilities, as it cannot be processed into a form that can be returned to the natural environment. The aftermath of coal's preparation method results in the formation of blackwater. The more coal is filtered and screened from rocks and minerals, the higher its efficiency and value.

Coal Slurry

Coal slurry is a fluid produced by washing coal with water and chemicals prior to shipping the coal to market. This waste product cannot be recycled, or broken down into usable substances, and is a major cause of concern, because storage procedures may not be environmentally sound over a long period of time. Coal slurry harms the marine life and aquatic animals, as well as causes air pollution. Components of coal slurry like carcinogenic compounds and heavy metals are permanently stored in toxic waste facilities because they are hardly transformed into a biodegradable natural form. Mankind acts of dumping the hazardous substance into a body of water or attempting to destroy the waste by incinerating it, all come on the account of harming the natural state of the environment, and the health of humans.

In order for coal to be ready to use, it has to go through a coal preparation plant. The extracted coal from underground or mountain formations goes through a process of screening. It is washed off from rocks, residues and sediments. The coal preparation plants are designed to yield a manageable quality of coal to meet the contract specifications and requirements that are set out by the standards and quality control commissions. A considerably clean lump of coal is tested for ash content, sulfuric level, calorific value and moisturization content. Coal goes through different circuit levels depending on the quality yielded. The main objective is to decrease the ash content and improve the maceral composition of the final clean product. Coal that is used for heating and energy is usually not thoroughly assessed, while coal that is used for electricity goes through an intensified line of purification. Despite the coal's contribution as a rich source of energy and electricity, the accumulation of coal slurry pursues long-term harmful effects on the environment. It is a byproduct of coal mining and preparation that is in a form of solid waste, typically composed of dirt, sludge, and rocks. The composition of coal slurry makes it almost impossible to decompose or biodegrade. Because the nature of the waste is ineffective as a reusable form of energy, it is used to mold a dam between two mountains, and then the slurry is placed behind the dam and impounded in the ground. The mixture of coal slurry and water creates liquid, solid and hazardous wastes that seep into the ground, evaporate into the air or are dumped into bodies of water. Coal slurry contains a large number of highly abrasive and corrosive minerals that have been washed and leached out of coal and rocks. The slurry also includes the infused chemicals that were used in the preparation cycle making it a viscous form of hazardous compounds. Some of these chemicals are acrylamide, butyl benzyl phthalate, hexachlorobenzene, naphthalene, chlorophenyl phenyl ether, and dichlorobenzidine. Heavy metals that are included in coal slurry are mercury, arsenic, lead, and nickel.

Since coal slurry is difficult to store, or dissociate, it is dumped into water systems by pipelines and progressive cavity pumps. The lethal coal slurry turns the water into blackwater, causing discoloration and low dissolved oxygen levels (Hypoxia), which drastically damages animal life. Blackwater changes the natural temperature of the water and the carbon levels, making it difficult for aquatic life to flourish. Aquatic animals rely on oxygen to breathe (aerobic); a low level of dissolved oxygen or state of hypoxia kills fish and sea life. A large number of native fish die, and Murray crayfish, shrimps, and yabbies are observed to go near the surface or the shore to gasp any form of oxygen causing them to die. Blackwater also causes fish to leave their natural habitat and migrate to areas where oxygen is readily available, affecting the temporary food chain of that part of the water system. Migration can also make it hard for fish to find their food and nutrition niche causing them to slowly die. Increased levels of carbon in the water can also cause the formation of radioactive chemicals in the water causing a dysfunction in the reproductive system of fish, which results into mutant fish. Fish can also lose the ability to feed because of visual impairment or the ability to reproduce and lay eggs. The change in water temperature can cause the water to easily evaporate in warm seasons, which decreases the water level and affects the sea life of aquatic animals and water plants that are always near the surface. High temperatures of water can make the environment favorable for bacterial growth like Escherichia coli, which can be found in fish and later affect humans who consume that fish. Blackwater in water bodies is also considered as a large threat to the irrigation and agricultural systems, as water used for irrigation can be used to water fruits, vegetables, and wheat. Blackwater can also seep into the soil, reducing soil nutrients that are an essential key for living organisms. The secondary effect happens when humans consume those blackwater-affected products, and get sick.

Coal power plants are a leading pollutant of carbon dioxide; they also contribute to the existence of smog, acid rain and toxic air pollution. Some of the most hazardous air toxins emitted from a coal power plant are sulfur dioxide, nitric oxide, particulate matter, and mercury. The emissions created from Coal plants create acidic particulates that damage the human health by penetrating the lung and circulating the blood stream, which causes the blood to acidify and oxidize, leading to death in very short periods of time. Other health aftermaths include chronic respiratory diseases, chronic bronchitis and aggravated asthma. The chemicals are also emitted into the air by the evaporation of blackwater in water and river systems. Once the water is evaporated, all the chemical compounds within the coal slurry are invited into the air, causing air pollution and putting risk on those who inhale it or live by the water system. It also pollutes the water, making it unsafe and hazardous for humans.

One of the riskiest components of blackwater to human health is carcinogens. Being exposed to coal slurry or blackwater can be deadly. Whether you are in close contact with a water system that is victimized by blackwater or drank water from your tap coming from a blackwater-contaminated resource; there is a high possibility of being exposed to carcinogen. Carcinogen is a radioactive agent that directly causes cancer, kidney failure, high blood pressure, miscarriages, and birth defects. One of the critical diseases that carcinogens create is the damaging of tissue cells, which damages the genome structure of the cellular metabolic process. Carcinogens are also reducing safe drinking water stains, which make it harder for humans to obtain a safe-drinking source.

It is often difficult to find alternative solutions for blackwater, because it results from coal slurry, which is non-biodegradable and hard to break its components into ones that can be used in their natural form in the environment. One of the possible alternatives will be using a semi-permeable membrane that filters the coal from rocks and residues without letting blackwater pass through the membranes and into the pipelines. Another alternative would be to manage the water systems, and use water turbines to circulate the streams of water, so blackwater does not concentrate in a certain area to the extent it leaves a damaging aftermath. An additional alternative might be adding chemical additives to reduce the hazardous effects of blackwater but that will also come on the account of marine life. The best alternative would be applying government tradable permits on the amount of blackwater that a coal preparation power plant can create, which will act as an incentive for mining sites to reduce any form of pollution, in fear of paying pollution taxes. Funding research on preventing blackwater spills and blackwater pollution is significant in order to take practical, preventative measures with a relatively low cost and the potential of being highly effective. And as always, the most important aspect is to advocate humanitarians and environmentalists on raising awareness for the current generation and the ones to come.

Greywater

Greywater is described as gently used waste water that comes from sources such as the bathroom, kitchen and laundry. It is relatively clean and cannot be compared with the water from toilets (blackwater). Generally, any water that is drained from the house other than toilet water can be described as greywater. It may also contain traced of food, grease, hair, dirt or particular household cleaning products.

Greywater is much easier to treat and recycle when compared with blackwater because there is no faecal matter that is a haven for harmful bacteria and disease causing pathogens.

If recycled properly, greywater can save approximately 70 litres of potable water per person per day in domestic households, therefore greywater recycling is one of a number of water solutions that we should look to in order to decrease our usage.

If your house is metered, recycling greywater can significantly reduce the volume of water you use thereby saving you money on your water bills.

Working of Greywater Recycling

Please remember that greywater is never going to be safe to drink, even when treated. However the reclaimed greywater can be used to flushing toilets, wash clothes and water your garden.

There are various ways to treat greywater, ranging from very simple methods to complex fully automated treatment systems, depending on what you want to use the water for.

Each of these is described in detail below.

Direct use Systems for Watering Plants

If you don't want to treat the water, you need to use it very quickly since bacteria present in the water feed on any organic matter present (e.g. skin particles, hair and detergents) and multiply very quickly. Once the bacteria have used up all the oxygen it will become foul smelling, and these anaerobic bacteria can contain harmful human pathogens.

Using this greywater is relatively simple, you can either syphon it directly from the bath or sink or you can fit a valve to the external waste pipe allowing you to direct the water to a water butt, so it can be used as needed to water the flowerbed.

It is not recommended to use greywater on fruit or crops, since there is a chance the plant will ingest the harmful pathogens when they are watered.

Biological Systems Non-food Debris

Sand Filter Method

The initial filtration can be simply achieved using a sand filter that removes any large particles. The greywater enters the top of the sand filter and travels down through the sand via gravity, with the sand removing any sizeable particles.

Once the greywater has been pre-treated, it can be filtered using a very simple soilbox consisting of four layers of material. The top layer is roughly 2 feet deep of humus-rich top soil, which sits on a bed of very fine building sand, which in turn sits on a layer of course sand. Finally there is a layer of pea-shingle at the bottom to achieve excellent drainage. The water is initially pumped in at the top of the soilbox, where it travels down via gravity through the 4 levels. Most of the filtration takes place in the topsoil level where soil organisms feed and reproduce using the nutrients in the soil, essentially purifying it.

Wetland

Another way of treating the greywater is using a wetland, where the water is retained at a level close to the surface, allowing aquatic plants such as reeds and bulrushes to flourish. Subsurface wetlands are considered better for treating greywater, as it lowers the chance of odours escaping, there is less chance of freezing during cold weather and has lower human contact, which is potentially dangerous. Bacteria, both aerobic and anaerobic treat the greywater. In addition, the roots of the plants absorb dissolved organic materials, helping speed up the process.

Biological Systems Including Food Debris

Greywater with food debris in it needs to be treated anaerobically using a septic tank (see treating domestic sewage for details on how a septic tank works). The water that comes out of the other end of the septic tank can then be treated with the soilbox filter or using wetlands as mentioned above.

Mechanical Filters for using the Water to Flush Toilets

It is a relatively simple process to undo the U-bend under a sink and capture wastewater in a bucket, which you could then manually pour into the cistern of a toilet. However practically, this is quite a time consuming job for something that can be easily automated.

Ecoplay is one of a number of companies that has produced a greywater pump that pumps water away from a variety of greywater sources including washing machines, showers and utility sinks. When the water enters the pump unit, it can then pump the water vertically to where it is required. The water is then treated in a storage tank (normally chemically by adding chlorine) before it is sent to the cistern of a toilet or the washing machine.

The greywater can also be treated using more expensive mechanical filters, such as microfiltration systems using membranes. We have looked at this in more detail in harvesting rainwater – so please look there for more information.

Uses for Treated Greywater

Using Greywater for Flushing Toilets

Typically, about a third of household water is used for flushing the toilet, but reclaimed greywater can be used to fulfil this purpose saving valuable potable mains water.

Once the greywater was gone through the complete filter process and treated with chemicals to kill all microorganisms it can be pumped from source back to the toilet via a header tank, and used as appropriate.

Using Greywater for Watering Plants

A great deal of water is also used in the summer to water plants – this is again a waste of potable water. This can be a major issue where we have droughts, which was experienced quite recently, back in early 2012.

The greywater can be used in combination with an irrigation system to water the garden automatically. This is good to have in place because it targets specific areas of your garden so it is a more environmentally friendly way to care for it.

Home water irrigation systems vary in complexity, with companies offering whole systems to maximise efficiency by very specific targeting.This water does not need to be treated with chemicals as any organic material remaining in the water can be absorbed by the plants. It is not recommended to use this type of water with home grown vegetables.

Using Greywater to Wash Clothes

Reclaimed greywater can also be used for washing clothes; again this has to be treated to a similar level as the water used for toilets. And like the toilet greywater, plumbing needs to be put in place to redirect the water to the appropriate place in the house.

Different Ways to Collect Greywater

1. Manual Collection

Greywater is collected from laundry, kitchen or bath water. The systems for collecting greywater vary from manual to completely automated with the manual systems being the simplest to use and the ones with the least costs of maintenance. The manual systems require nothing more than a bucket and jug for collecting greywater by hand. The water can then be used either for irrigation plants or as a substitute for using fresh water for flushing.

2. Simple Piping

The second more complex system creates an automated connection between laundry and the landscape. This system uses plumbing connections between washing machine's waste water drain pipe and the lawn. Here, the water flows via underground piping into the yard where it is used to irrigate the land. It has a low cost of materials and requires some maintenance to ensure that the plumbing does not become clogged up with residue from the soapy water. This kind of system can be built by oneself with little effort and training plus the materials required such as hose pipes and connectors.

3. Integrated Sewage and Piping

The third system involves integrating the whole houses plumbing into a greywater collection system. These systems are the most expensive and the most maintenance intensive options available. They also require that one hires professionals in order to ensure that the work is done properly. The systems usually connect bath, laundry and kitchen drains to a system that dispenses greywater into a collection tank. The water usually needs to be used within 24 hours of collection in order to avoid bacteria from growing to numbers that cause foul smells. The water can however be treated with chlorine or iodine to allow for longer storage.

Some systems can even allow for the merging of greywater to toilet plumbing ensuring that one does not flush the toilet with fresh water. The systems can be further integrated to include greywater outlets that merge into the existing sewage systems to ensure there is no overflow. Some systems can also come with an emergency shut of switch to ensure that in case the system fails, there is no backflow into the main freshwater piping. The amount of piping and pumping involved in these most complex systems means they are more susceptible to blocking due to residual soap. They therefore need more maintenance than other systems.

Benefits of using GreyWater

1. Fresh water Conservation

Use of greywater reduces the amount of fresh water used up in the household for other functions. The reusing of greywater to flush toilets and water plants reduces the load of fresh water required and if enough people make the same choice, the water demands can be reduced so drastically as to positively impact the environment. This is particularly important in regions experiencing dry climates or going through droughts.

2. Reduction of water wastage

The use of freshwater for all the different functions in a house produce lots of waste water that all ends up flowing into the sewer systems. This creates unnecessary wastage. By reusing greywater for one other function before it is dumped into the sewage, the amount of water wasted can be halved. Thus, it means fewer resources can be diverted to sewage treatment and that the saving of even an extra liter of water which could have been used before its disposal increases as well, thus increasing water use efficiency.

3. Reduced energy consumption

With reduced fresh water demands for each household, there is reduced energy demand required to pump the water into the house. Secondly, water reuse reduces the load that treatment centers have to handle in ensuring that it is purified thus reducing total energy required by both water distribution and sewage purifying companies. Reduced energy use means less electricity and in turn, fossil fuel use is also cut down eventually helping to reduce greenhouse gas emissions. It means that energy can be diverted into other resource intensive requirements or stored for later use.

4. Reduced chemical consumption

The reduction of water sent to sewage plants also leads to reduced levels in the amount of chemicals used in treating the water. With a smaller load, sewage treatment plants do not require as much use of chemicals which is beneficial to the environment as it reduces both the cost of sewage treatment as well as reducing the demand for chemical production that goes to benefit the environment. It also reduces the chance of accidental dumping of chemical waste by sewage processing companies. Greywater use also reduces the need to fertilize the yards with chemical fertilizers due to the nutrients the water already contains.

5. Beautification of landscapes

The use of greywater to water yards provides an alternative source of water to plants, especially in regions where plants lack it. Regions that use reclaimed greywater thus provide alternative use for water that would have gone to waste. Therefore, it creates beautiful landscapes as plants maintain greenness and bloom, even in regions where the climates are harsh. Coupled with other technologies such as the use of mulch and drip irrigation, the water provided can last longer to ensure the beautiful state of one's yard is maintained even during the driest of seasons.

6. Reuse of nutrients

The use of greywater reintroduces nutrients that would have otherwise been lost to the sewage system. Kitchen and bath water contain a lot of organic material that is not harmful to plants. This is unlike blackwater where the organic levels are too high. Blackwater also has lots of harmful bacteria e.g. *E. coli* that could cause disease if they contaminate the soil. The greywater provides plants with good nutrients that go towards increased beautification.

7. Organic filtering of water

The greywater used in watering plants is purified by both plant and soil action which negates the use of chemicals that might, in excess, harm the soil and the organisms that live within it. Organic filtering serves to ensure the environments safety.

8. Groundwater recharge

The greywater used often ends up being filtered by the soil as it moves down and ends up in the ground water where is replenishes the levels therein. This is a good thing as it ensures waterways such as rivers sourcing their water from the underground aquifers are always fed.

Yellow Water

This yellow water is specifically Urine collected with some particular Channels and it is not contaminated either with greywater or BlackWater. So now we go to Sources of WasteWater.

Sewage

Sewage, also called wastewater, is the contaminated water from homes, schools, and businesses. It comes from toilets, showers, clothes washers, dishwashers, etc. The contaminants include fecal matter, urine, soaps, detergents, food particles, hair, rags, paper, toys, dead goldfish, and anything else that is disposed in a drain. A person creates an average of 60 to 100 gallons of wastewater every day. Sewers are a network of pipes that bring the sewage to the treatment plant for treatment. Treatment is the continual process of removing the contaminants from the wastewater and then processing the removed contaminants into a product that can be safely recycled.

There are two types of sewage: treated and untreated.

Treated Sewage

Treated sewage refers to wastewater or sewage which has passed through a treatment plant. Sewage goes through several stages in the treatment process ensuring that all harmful bacteria, pollutants and contaminants are eliminated. The stages of sewage treatment include pre-treatment, primary, secondary and tertiary sewage treatment. The last stage usually involves the use of UV light to ensure all bacteria and/or viruses are removed. After treatment, the water will usually pass into rivers or seas or be reused for irrigation and agricultural purposes.

Coming into contact with treated sewage is rare, but can arise if large flooding events affect sewage treatment plants or in very rare cases there is a pump or other equipment failure that results in wastewater spreading over land and potentially into property.

Untreated Sewage

Untreated sewage refers to wastewater which contains harmful waterborne pathogens and bacteria and which has not yet gone through a sewage treatment plant. Raw sewage originates from broken toilet pipes, overspills, industry leakages and heavy storms. It poses an extremely high risk to human and animal health and the longer it sits and stagnates a home or a business, the greater amount of bacteria it will contain.

Quite often, in the poorer areas of the world, sewage can get dumped anywhere and this is unfortunate as people have no access to proper treatment plants, therefore increasing health risks in those areas.

When a flood occurs, more often than not, the water contains untreated sewage and it therefore must be dealt with straight away. CleanSafe Services respond immediately and we can get to you within less than 2 hours, reducing business interruption and majorly decreasing the risk of damage to your health and your property.

Pollutants

Organic pollutants and Nutrients

Sewage is a complex mixture of chemicals, with many distinctive chemical characteristics. These include high concentrations of ammonium, nitrate, nitrogen, phosphorus, high conductivity (due to high dissolved solids), high alkalinity, with pH typically ranging between 7 and 8. The organic matter of sewage is measured by determining its biological oxygen demand (BOD) or the chemical oxygen demand (COD).

Pathogens

Sewage contains human feces, and therefore often contains pathogens of one of the four types:

- Bacteria (for example Salmonella, Shigella, Campylobacter, Vibrio cholerae),

- Viruses (for example hepatitis A, rotavirus, enteroviruses),

- Protozoa (for example Entamoeba histolytica, Giardia lamblia, Cryptosporidium parvum) and

- Parasites such as helminths and their eggs (e.g. ascaris (roundworm), ancylostoma (hookworm), trichuris (whipworm)).

Sewage can be monitored for both disease-causing and benign organisms with a variety of techniques. Traditional techniques involve filtering, staining, and examining samples under a microscope. Much more sensitive and specific testing can be accomplished with DNA sequencing, such as when looking for rare organisms, attempting eradication, testing specifically for drug-resistant strains, or discovering new species. Sequencing DNA from an environmental sample is known as metagenomics.

Micro-pollutants

Sewage also contains environmental persistent pharmaceutical pollutants. Trihalomethanes can also be present as a result of past disinfection.

Sewage has also been analyzed to determine relative rates of use of prescription and illegal drugs among municipal populations.

Health and Environmental Aspects

All categories of sewage are likely to carry pathogenic organisms that can transmit disease to humans and animals. Sewage also contains organic matter that can cause odor and attract flies.

Sewage contains nutrients that may cause eutrophication of receiving water bodies; and can lead to ecotoxicity.

Collection

A medieval waste pipe in Stockholm Old Town formerly deposited sewage on the street to be flushed away by rain.

Sewage canal of a medieval house as depicted *St. Barbara Altarpiece* in the National Museum in Warsaw.

A system of sewer pipes (sewers) collects sewage and takes it for treatment or disposal. The system of sewers is called *sewerage* or *sewerage system* in British English and *sewage system* in American English. Where a main sewerage system has not been provided, sewage may be collected from homes by pipes into septic tanks or cesspits, where it may be treated or collected in vehicles and taken for treatment or disposal. Properly functioning septic tanks require emptying every 2–5 years depending on the load of the system.

Treatment

Sewage treatment is the process of removing the contaminants from sewage to produce liquid and solid (sludge) suitable for discharge to the environment or for reuse. It is a form of waste management. A septic tank or other on-site wastewater treatment system such as biofilters or constructed wetlands can be used to treat sewage close to where it is created.

Sewage treatment results in sewage sludge which requires sewage sludge treatment before safe disposal or reuse. Under certain circumstances, the treated sewage sludge might be termed "biosolids" and can be used as a fertilizer.

In developed countries sewage collection and treatment is typically subject to local and national regulations and standards.

Disposal

Raw sewage is also disposed of to rivers, streams, and the sea in many parts of the world. Doing so can lead to serious pollution of the receiving water. This is common in developing countries and may still occur in some developed countries, for various reasons - usually related to costs.

Reuse of Treated or Untreated Sewage

Increasingly, agriculture is using untreated wastewater for irrigation. Cities provide lucrative markets for fresh produce, so are attractive to farmers. However, because agriculture has to compete for increasingly scarce water resources with industry and municipal users, there is often no alternative for farmers but to use water polluted with urban waste, including sewage, directly to water their crops. There can be significant health hazards related to using water loaded with pathogens in this way, especially if people eat raw vegetables that have been irrigated with the polluted water.

The International Water Management Institute has worked in India, Pakistan, Vietnam, Ghana, Ethiopia, Mexico and other countries on various projects aimed at assessing and reducing risks of wastewater irrigation. They advocate a 'multiple-barrier' approach to wastewater use, where farmers are encouraged to adopt various risk-reducing behaviours. These include ceasing irrigation a few days before harvesting to allow pathogens to die off in the sunlight, applying water carefully so it does not contaminate leaves likely to be eaten raw, cleaning vegetables with disinfectant or allowing fecal sludge used in farming to dry before being used as a human manure. The World Health Organization has developed guidelines for safe water use.

Legislation

European Union

Council Directive 91/271/EEC on Urban Wastewater Treatment was adopted on 21 May 1991, amended by the Commission Directive 98/15/EC. Commission Decision 93/481/EEC defines the information that Member States should provide the Commission on the state of implementation of the Directive.

Industrial Wastewater

Industrial wastewater is generated as a consequence of industrial activities. Industrial wastewater is generated as a result of industrial activity. There is a wide range of types of industrial wastewater (e.g. from processes, cleaning and cooling), with different types of pollutants.

Most industrial processes use water in one way or another. Once used, the water has to be treated before being disposed of, regardless of whether it is returned to the natural environment or into the sewage network.

For the former, the treatment must be sufficient so that the discharge has no detrimental environmental impact; and, if discharged into the sewage network, the wastewater physical and chemical properties must comply with current regulations. There is a third option for already treated industrial wastewater: re-use.

Since water is a natural resource that should not be wasted, the most sustainable alternative is to treat wastewater until its quality is appropriate for re-use in the process. Environmental regulations, which are increasingly demanding, mean that re-use is the most competitive option in many cases.

At Condorchem Envitech, we work to optimize the use of water in industry, promoting its re-use after the production processes via zero discharge technologies, provided that these are economically and environmentally viable.

If zero discharge is not possible, we treat it until it is within the legally established discharge limits to avoid sanctions or other serious financial or social repercussions and loss of image.

Industrial wastewater with significant carbon loading that is treated under intended or unintended anaerobic conditions will produce CH_4. Assessment of CH_4 production potential from industrial wastewater streams is based on the concentration of degradable organic matter in the wastewater, the volume of wastewater, and the propensity of the industrial sector to treat their wastewater in anaerobic systems. Using these criteria, major industrial wastewater sources with high CH_4 gas production potential can be identified as follows:

- pulp and paper manufacture

- meat and poultry processing (slaughter houses)

- alcohol, beer, starch production

- organic chemicals production

- other food and drink processing (dairy products, vegetable oil, fruits and vegetables, canneries, juice making, etc.)

Both the pulp and paper industry and the meat and poultry processing industries produce large volumes of wastewater that contain high levels of degradable organics. The meat and poultry processing facilities typically employ anaerobic lagoons to treat their wastewater, while the paper and pulp industry also use lagoons and anaerobic reactors. The nonanimal food and beverage industries produce considerable amounts of wastewater with significant organic carbon levels and are also known to use anaerobic processes such as lagoons and anaerobic reactors. Anaerobic reactors treating industrial effluents with biogas facilities are usually linked with recovery of the generated CH_4 for energy. The development of emission factors and activity data is more complex because there are many types of wastewater, and many different industries to track. The most accurate estimates of emissions for this source category would be based on measured data from point sources. Due to the high costs of measurements and the potentially large number of point sources, collecting comprehensive measurement data is very difficult.

Total CH_4 emissions from industrial wastewater are calculated as follows:

$$CH_4 \text{ emissions} = \sum i[(TOWi - Si)EFi - Ri]$$

where CH_4 emissions = CH_4 emissions in inventory year, kg CH_4 per year, TOW_i = total organically degradable material in wastewater from industry I in inventory year, kg COD per year, i = industrial sector, S_i = organic component removed as sludge in inventory year, kg COD per year, and EF_i = emission factor for industry i, kg CH_4/kg COD for treatment/discharge pathway or system(s) used in inventory year. If more than one treatment practice is used in an industry, this factor would need to be a weighted average. R_i = amount of CH_4 recovered in inventory year, kg CH_4 per year.

There are significant differences in the CH_4-emitting potential of different types of industrial wastewater. To the extent possible, data should be collected to determine the maximum CH_4-producing capacity (Bo) in each industry. As mentioned before, the methane correction factor (MCF) indicates the extent to which the CH_4-producing potential (Bo) is realized in each type of treatment method.

CH_4 emission factor for industrial wastewater is calculated as follows:

$$EFj = Bo \cdot MCFj$$

where EF_j = emission factor for each treatment/discharge pathway or system, kg CH_4/kg COD, j = each treatment/discharge pathway or system, B_o = maximum CH_4 producing capacity, kg CH_4/kg COD, and MCF_j = methane correction factor (fraction).

Default MCF values, which are based on expert judgment as given in IPCC report, 2006.

Industrial production data and wastewater outflows may be obtained from national statistics, regulatory agencies, wastewater treatment associations, or industry associations.

Similar estimates have to be brought out for oxides of nitrogen. Nitrous oxide (N_2O) emissions can occur as direct emissions from treatment plants or from indirect emissions from wastewater after disposal of effluent into waterways, lakes, or the sea. Direct emissions from nitrification and denitrification at wastewater treatment plants may be considered as a minor source. The simplified general equation is as follows:

N_2O emissions from wastewater effluent

$$N_2O \text{ emissions} = N_{efficient} \cdot EF_{effluent} \cdot 44/28$$

where N_2O emissions = N_2O emissions in inventory year, kg N_2O per year, $N_{effluent}$ = nitrogen in the effluent discharged to aquatic environments, kg N per year, $EF_{effluent}$ = emission factor for N_2O emissions from discharged to wastewater, kg N_2O-N/kg N, and the factor 44/28 is the conversion of kg N_2O-N into kg N_2O.

Total Nitrogen in the Effluent is Calculated as Follows

$$NE_{EFFLUENT} = (P \cdot Protein \cdot F_{NPR} \cdot F_{NON-CON} \cdot F_{IND-COM}) - N_{SLUDGE}$$

where $NE_{EFFLUENT}$ = total annual amount of nitrogen in the wastewater effluent, kg N per year, P = human population, Protein = annual per capita protein consumption, kg/person per year, F_{NPR} = fraction of nitrogen in protein, default = 0.16, kg N/kg protein, $F_{NON-CON}$ = factor for nonconsumed protein added to the wastewater, $F_{IND-COM}$ = factor for industrial and commercial co-discharged protein into the sewer system, and N_{SLUDGE} = nitrogen removed with sludge (default = zero), kg N per year.

N_2O emission from centralized wastewater treatment processes is as follows:

$$N_2O_{PLANTS} = P \cdot T_{PLANT} \cdot F_{IND-COM} \cdot EF_{PLANT}$$

where N_2O_{PLANTS} = total N_2O emissions from plants in inventory year, kg N_2O per year, P = human population, T_{PLANT} = degree of utilization of modern, centralized WWT plants, %, $F_{IND-COM}$ = fraction of industrial and commercial co-discharged protein (default = 1.25, based on data in and expert judgment), and EF_{PLANT} = emission factor, 3.2 g N_2O/person per year.

Sources of Industrial Wastewater

Almost all industries produce some form of wastewater. Equipment designed for solid-liquid separation, such as a filter press, provides for the reuse of water and cost-effective disposal of dewatered solids.

Pulp and Paper

Paper manufacturing is an energy-and water-intensive industry. From biochemical oxygen demand (BOD) to chemical oxygen demand (COD), solids to chlorinated organic compounds, the pulp and paper industry generates a high volume of wastewater.

Mines and Quarries

Mining of metallic ores and precious metals can produce wastewater contaminated by rock particles, metals, acids and salts, as well as hydraulic oils.

M.W. Watermark 1200mm filter press. Built for a neutralized mine wastewater application.

Oil and Gas

Hydraulic fracturing, or fracking, produces large volumes of wastewater as a process for extracting shale gas. The wastewater may contain high concentrations of dissolved solids, radionuclides and metals.

M.W. Watermark 1200mm trailer-mounted filter press. Built for produced water recovery.

Iron and Steel

The most common pollutants in the iron and steel sector are BOD, COD, oil, metals, acids, phenols and cyanide.

M.W. Watermark 1200mm filter press manufactured for sand filter backwash dewatering.

Food Industry

Wastewater produced as a by-product of the food industry is often biodegradable and nontoxic. However, it may contain high concentrations of BOD, suspended solids (SS) and even pesticides.

M.W. Watermark 800mm filter press with a semi-automatic plate shifter. Application: Egg production wastewater.

Chemicals

Using complex organic chemicals, the chemicals industry can generate wastewater contaminated by COD, organic chemicals, heavy metals, SS and cyanide.

References

- World Health Organization (2006). Guidelines for the safe use of wastewater, excreta, and greywater. World Health Organization. p. 31. ISBN 9241546859. OCLC 71253096

- van der Baan, Mirko; Calixto, Frank J. (2017-07-01). "Human-induced seismicity and large-scale hydro-carbon production in the USA and Canada". Geochemistry, Geophysics, Geosystems. 18(7): 2467–2485. doi:10.1002/2017gc006915. ISSN 1525-2027

- "Environmental policy and legislation". Department of Environmental and Heritage Protection. Queensland Government. Archived from the original on 20 October 2017. Retrieved 20 October 2017

- Tilley, E.; Ulrich, L.; Lüthi, C.; Reymond, Ph.; Zurbrügg, C. Compendium of Sanitation Systems and Technologies (2nd Revised ed.). Duebendorf, Switzerland: Swiss Federal Institute of Aquatic Science and Technology (Eawag). p. 10. ISBN 978-3-906484-57-0

- "An Act Providing For A Comprehensive Water Quality Management And For Other Purposes". The LawPhil Project. Archived from the original on 21 September 2016. Retrieved 30 September 2016

Chapter 2

Wastewater Quality Indicators

Wastewater quality indicators are the tests and techniques implemented for assessing the suitability of wastewater for re-use or disposal. These tests measure biological, chemical and physical characteristics of wastewater. This chapter has been carefully written to provide an easy understanding of the varied characteristics that are measured by such quality indicators, like biochemical oxygen demand, chemical oxygen demand, total dissolved solids, total suspended solids, etc.

Wastewater quality indicators such as the biochemical oxygen demand (BOD) and the chemical oxygen demand(COD) are essentially laboratory tests to determine whether or not a specific wastewater will have a significantadverse effect upon fish or upon aquatic plant life.

Wastewater Biochemical Oxygen Demand and Chemical Oxygen Demand

Any oxidizable material present in a natural waterway or in an industrial wastewater will be oxidized both bybiochemical (bacterial) or chemical processes. The result is that the oxygen content of the water will be decreased. Basically, the reaction for biochemical oxidation may be written as:

:Oxidizable material + bacteria + nutrient + $O_2 \rightarrow CO_2 + H_2O$ + oxidized inorganics such as NO_3 or SO_4

Oxygen consumption by reducing chemicals such as sulfides and nitrites is typified as follows:

:$S^{--} + 2\,O_2 \rightarrow SO_4^{-}$

:$NO_2^{-} + \frac{1}{2}\,O_2 \rightarrow NO_3^{-}$

Since all natural waterways contain bacteria and nutrient, almost any waste compounds introduced into suchwaterways will initiate biochemical reactions (such as shown above). Those biochemical reactions create what ismeasured in the laboratory as the Biochemical Oxygen Demand (BOD).

Oxidizable chemicals (such as reducing chemicals) introduced into a natural water will similarly initiate chemicalreactions (such as shown above). Those chemical reactions create what is measured in the laboratory as theChemical Oxygen Demand (COD).

Both the BOD and COD tests are a measure of the relative oxygen-depletion effect of a waste contaminant. Bothhave been widely adopted as a measure of pollution effect. The BOD test measures the oxygen demand ofbiodegradable pollutants whereas the COD test measures the oxygen demand of biogradable pollutants plus theoxygen demand of non-biodegradable oxidizable pollutants.

The so-called 5-day BOD measures the amount of oxygen consumed by biochemical oxidation of waste contaminantsin a 5-day period. The total amount of oxygen consumed when the biochemical reaction is allowed to proceed tocompletion is called the Ultimate BOD. The Ultimate BOD is too time consuming, so the 5-day BOD has almostuniversally been adopted as a measure of relative pollution effect.

There are also many different COD tests. Perhaps, the most common is the 4-hour COD.

It should be emphasized that there is no generalized correlation between the 5-day BOD and the Ultimate BOD. Likewise, there is no generalized correlation between BOD and COD. It is possible to develop such correlations for aspecific waste contaminant in a specific wastewater stream, but such correlations cannot be generalized for use withany other waste contaminants or wastewater streams.

Physical Characteristics

Temperature

Aquatic organisms cannot survive outside of specific temperature ranges. Irrigation runoff and water cooling of power stations may elevate temperatures above the acceptable range for some species. Elevated temperature can also cause an algae bloom which reduces oxygen levels. Temperature may be measured with a calibrated thermometer.

Solids

Solid material in wastewater may be dissolved, suspended, or settled. Total dissolved solids or TDS (sometimes called filterable residue) is measured as the mass of residue remaining when a measured volume of filtered water is evaporated. The mass of dried solids remaining on the filter is called total suspended solids (TSS) or nonfiltrable residue.Settleable solids are measured as the visible volume accumulated at the bottom of an Imhoff cone after water has settled for one hour. Turbidity is a measure of the light scattering ability of suspended matter in the water. Salinity measures water density or conductivity changes caused by dissolved materials.

Chemical Characteristics

Virtually any chemical may be found in water, but routine testing is commonly limited to a few chemical elements of unique significance.

Hydrogen

Water ionizes into hydronium (H_3O) cations and hydroxyl (OH) anions. The concentration of ionized hydrogen (as protonated water) is expressed as pH.

Nitrogen

Nitrogen is an important nutrient for plant and animal growth. Atmospheric nitrogen is less biologically available than dissolved nitrogen in the form of ammonia and nitrates. Availability of dissolved nitrogen may contribute to algal blooms. Ammonia and organic forms of nitrogen are often measured as Total Kjeldahl Nitrogen, and analysis for inorganic forms of nitrogen may be performed for more accurate estimates of total nitrogen content.

Phosphates

Total Phosphorus and Phosphate, PO_4^{-3}

Phosphates enter the water ways through both non-point sources and point sources. Non-point source (NPS) pollution refers to water pollution from diffuse sources. Nonpoint source pollution can be contrasted with point source pollution, where discharges occur to a body of water at a single location. The non-point sources of phosphates include: natural decomposition of rocks and minerals, storm water runoff, agricultural runoff, erosion and sedimentation, atmospheric deposition, and direct input by animals/wildlife; whereas: point sources may include: waste water treatment plants and permitted industrial discharges. In general, the non-point source pollution typically is significantly higher than the point sources of pollution. Therefore, the key to sound management is to limit the input from both point and non-point sources of phosphate. High concentration of phosphate in water bodies is an indication of pollution and largely responsible for eutrophication.

Phosphates are not toxic to people or animals unless they are present in very high levels. Digestive problems could occur from extremely high levels of phosphate.

The following criteria for total phosphorus were recommended by the U.S. Environmental Protection Agency.

1. No more than 0.1 mg/L for streams which do not empty into reservoirs,

2. No more than 0.05 mg/L for streams discharging into reservoirs, and

3. No more than 0.025 mg/L for reservoirs.

Phosphorus is normally low (< 1 mg/l) in clean potable water sources and usually not regulated.

Chlorine

Chlorine has been widely used for bleaching, as a disinfectant, and for biofouling prevention in water cooling systems. Remaining concentrations of oxidizing hypochlorous acid and hypochlorite ions may be measured as chlorine residual to estimate effectiveness of disinfection or to demonstrate safety for discharge to aquatic ecosystems.

Biological Characteristics

Water may be tested by a bioassay comparing survival of an aquatic test species in the wastewater in comparison to water from some other source. Water may also be evaluated to determine the approximate biological population of the wastewater. Pathogenic micro-organisms using water as a means of moving from one host to another may be present in sewage. Coliform index measures the population of an organism commonly found in the intestines of warm-blooded animals as an indicator of the possible presence of other intestinal pathogens.

Biochemical Oxygen Demand

Biochemical oxygen demand or BOD is a chemical procedure for determining the amount of dissolved oxygen needed by aerobic biological organisms in a body of water to break down organic material present in a given water sample at certain temperature over a specific time period. It is not a precise quantitative test, although it is widely used as an indication of the organic quality of water. It is most commonly expressed in milligrams of oxygen consumed per liter of sample during 5 days (BOD_5) of incubation at 20°C and is often used as a robust surrogate of the degree of organic pollution of water.

BOD directly affects the amount of dissolved oxygen in rivers and streams. The rate of oxygen consumption is affected by a number of variables: temperature, pH, the presence of certain kinds of microorganisms, and the type of organic and inorganic material in the water.

The greater the BOD, the more rapidly oxygen is depleted in the stream. This means less oxygen is available to higher forms of aquatic life. The consequences of high BOD are the same as those for low dissolved oxygen: aquatic organisms become stressed, suffocate, and die.

Sources of BOD include topsoil, leaves and woody debris; animal manure; effluents from pulp and paper mills, wastewater treatment plants, feedlots, and food-processing plants; failing septic systems; and urban stormwater runoff.

BOD is affected by the same factors that affect dissolved oxygen. BOD measurement requires taking two measurements. One is measured immediately for dissolved oxygen (initial), and the second is incubated in the lab for 5 days and then tested for the amount of dissolved oxygen remaining (final). This represents the amount of oxygen consumed by microorganisms to break down the organic matter present in the sample during the incubation period.

BOD Measurement Problems

When encountering a problem with BOD values, most operators blame the instrument or probe first. After all, it's what gives you the values! There can be many factors leading to the problem from dirty BOD bottles, DI water, bad seed, bubbles, non-linearity and so on.

Calibration: Regardless of technology, it is recommended to only calibrate the instrument once a day before readings are taken. Make sure there are no water droplets on the probe tip before calibrating and that the probe is in a 100% water-saturated air environment. A BOD bottle with a little water in the bottom is all that's needed. Place the probe back in this bottle when not in use. Make sure this bottle is kept clean as shown in the bottle on the right of the image. The bottle on the left is not a clean environment and could affect the calibration.

Probe Care: Polarographic sensors must have fresh electrolyte and membranes along with a maintained anode and cathode for optimal performance. Optical sensors will need an occasional visual check of the paint layer on the sensor cap.

Warm-Up Time: Polarographic probes require sufficient warm-up time of 5-15 minutes before calibrating and using the probe. Improper warm-up time can easily lead to data drifting due to an inaccurate calibration. Optical probes do not require a warm-up period.

A variety of strategies are employed to deal with specific sample types. These include varying dilutions and diluent seeding. It is often desirable to distinguish between carbonaceous and nitrogenous demand, in which case a nitrification inhibitor is used. Toxic and chlorinated samples also need special handling. The operator must be familiar with standard methods and with the technical literautre on the subject.

When properly used, the BOD test provides a reliable characterization of wastewater. It can be expected to be a standard for regulatory agencies for many years even though its use as a control tool is limited by the 3 or 5 day wait required for the test (and sometimes 20 days!). Various methods (based on short-term monitoring and extrapolation) of quickly estimating the probable results of the BOD test on a sample have been devised.

Background

Most natural waters contain small quantities of organic compounds. Aquatic microorganisms have evolved to use some of these compounds as food. Microorganisms living in oxygenated waters use dissolved oxygen to oxidatively degrade the organic compounds, releasing energy which is used for growth and reproduction. Populations of these microorganisms tend to increase in proportion to the amount of food available. This microbial metabolism creates an oxygen demand proportional to the amount of organic compounds useful as food. Under some circumstances, microbial metabolism can consume dissolved oxygen faster than atmospheric oxygen can dissolve into the water or the autotrophic community (algae, cyanobacteria and macrophytes) can produce. Fish and aquatic insects may die when oxygen is depleted by microbial metabolism.

Biochemical oxygen demand is the amount of oxygen required for microbial metabolism of organic compounds in water. This demand occurs over some variable period of time depending on temperature, nutrient concentrations, and the enzymes available to indigenous microbial populations. The amount of oxygen required to completely oxidize the organic compounds to carbon dioxide and water through generations of microbial growth, death, decay, and cannibalism is total biochemical oxygen demand (total BOD). Total BOD is of more significance to food webs than to water quality. Dissolved oxygen depletion is most likely to become evident during the initial aquatic microbial population explosion in response to a large amount of organic material. If the microbial population deoxygenates the water, however, that lack of oxygen imposes a limit on population growth of aerobic aquatic microbial organisms resulting in a longer term food surplus and oxygen deficit.

A standard temperature at which BOD testing should be carried out was first proposed by the Royal Commission on Sewage Disposal in its eighth report in 1912:

"An effluent in order to comply with the general standard must not contain as discharged more than 3 parts per 100,000 of suspended matter, and with its suspended matters included must not take up at 65°F more than 2.0 parts per 100,000 of dissolved oxygen in 5 days. This general standard should be prescribed either by Statute or by order of the Central Authority, and should be subject to modifications by that Authority after an interval of not less than ten years.

This was later standardised at 68 °F and then 20 °C. This temperature may be significantly different from the temperature of the natural environment of the water being tested.

Although the Royal Commission on Sewage Disposal proposed 5 days as an adequate test period for rivers of the United Kingdom of Great Britain and Ireland, longer periods were investigated for North American rivers. Incubation periods of 1, 2, 5, 10 and 20 days were being used into the mid-20th century. Keeping dissolved oxygen available at their chosen temperature, investigators found up to 99 percent of total BOD was exerted within 20 days, 90 percent within 10 days, and approximately 68 percent within 5 days. Variable microbial population shifts to nitrifying bacteria limit test reproducibility for periods greater than 5 days. The 5-day test protocol with acceptably reproducible results emphasizing carbonaceous BOD has been endorsed by the United States Environmental Protection Agency. This 5-day BOD test result may be described as the amount of oxygen required for aquatic microorganisms to stabilize decomposable organic matter under aerobic conditions. Stabilization, in this context, may be perceived in general terms as the conversion of food to living aquatic fauna. Although these fauna will continue to exert biochemical oxygen demand as they die, that tends to occur within a more stable evolved ecosystem including higher trophic levels.

Taking samples from the influent raw wastewater stream for BOD measurements at a wastewater treatment plant in Haran-Al-Awamied near Damascus in Syria

Typical Values

Most pristine rivers will have a 5-day carbonaceous BOD below 1 mg/L. Moderately polluted rivers may have a BOD value in the range of 2 to 8 mg/L. Rivers may be considered severely polluted when BOD values exceed 8 mg/L. Municipal sewage that is efficiently treated by a three-stage process would have a value of about 20 mg/L or less. Untreated sewage varies, but averages around 600 mg/L in Europe and as low as 200 mg/L in the U.S., or where there is severe groundwater or surface water Infiltration/Inflow. The generally lower values in the U.S. derive from the much greater water use per capita than in other parts of the world.

Methods

There are two commonly recognized methods for the measurement of BOD.

Dilution Method

This standard method is recognized by U.S. EPA, which is labeled Method 5210B in the Standard Methods for the Examination of Water and Wastewater In order to obtain BOD_5, dissolved oxygen (DO) concentrations in a sample must be measured before and after the incubation period, and appropriately adjusted by the sample corresponding dilution factor. This analysis is performed using 300 ml incubation bottles in which buffered dilution water is dosed with seed microorganisms and stored for 5 days in the dark room at 20 °C to prevent DO production via photosynthesis. In addition to the various dilutions of BOD samples, this procedure requires dilution water blanks, glucose glutamic acid (GGA) controls, and seed controls. The dilution water blank is used to confirm the quality of the dilution water that is used to dilute the other samples. This is necessary because impurities in the dilution water may cause significant alterations in the results. The GGA control is a standardized solution to determine the quality of the seed, where its recommended BOD_5 concentration is 198 mg/l ± 30.5 mg/l. For measurement of carbonaceous BOD (cBOD), a nitrification inhibitor is added after the dilution water has been added to the sample. The inhibitor hinders the oxidation of ammonia nitrogen, which supplies the nitrogenous BOD (nBOD). When performing the BOD_5 test, it is conventional practice to measure only cBOD because nitrogenous demand does not reflect the oxygen demand from organic matter. This is because nBOD is generated by the breakdown of proteins, whereas cBOD is produced by the breakdown of organic molecules.

BOD_5 is calculated by:

$$\text{Unseeded: } BOD_5 = \frac{(D_0 - D_5)}{P}$$

$$\text{Seeded: } BOD_5 = \frac{(D_0 - D_5) - (B_0 - B_5)f}{P}$$

where:

D_0 is the dissolved oxygen (DO) of the diluted solution after preparation (mg/l)

D_5 is the DO of the diluted solution after 5 day incubation (mg/l)

P is the decimal dilution factor

B_0 is the DO of diluted seed sample after preparation (mg/l)

B_5 is the DO of diluted seed sample after 5 day incubation (mg/l)

f is the ratio of seed volume in dilution solution to seed volume in BOD test on seed

Manometric Method

This method is limited to the measurement of the oxygen consumption due only to carbonaceous oxidation. Ammonia oxidation is inhibited.

The sample is kept in a sealed container fitted with a pressure sensor. A substance that absorbs carbon dioxide (typically lithium hydroxide) is added in the container above the sample level. The sample is stored in conditions identical to the dilution method. Oxygen is consumed and, as ammonia oxidation is inhibited, carbon dioxide is released. The total amount of gas, and thus the pressure, decreases because carbon dioxide is absorbed. From the drop of pressure, the sensor electronics computes and displays the consumed quantity of oxygen.

The main advantages of this method compared to the dilution method are:

- simplicity: no dilution of sample required, no seeding, no blank sample.

- direct reading of BOD value.

- continuous display of BOD value at the current incubation time.

Alternative Methods

Biosensor

An alternative to measure BOD is the development of biosensors, which are devices for the detection of an analyte that combines a biological component with a physicochemical detector component. Enzymes are the most widely used biological sensing elements in the fabrication of biosensors. Their application in biosensor construction is limited by the tedious, time consuming and costly enzyme purification methods. Microorganisms provide an ideal alternative to these bottlenecks.

The vast variety of micro organisms are relatively easy to maintain in pure cultures, grow and harvest at low cost. Moreover, the use of microbes in biosensor field has opened up new possibilities and advantages such as ease of handling, preparation and low cost of device. A number of pure cultures, e.g. *Trichosporon cutaneum, Bacillus cereus, Klebsiella oxytoca, Pseudomonas sp.* etc. individually, have been used by many workers for the construction of BOD biosensor. On the other hand, many workers have immobilized activated sludge, or a mixture of two or three bacterial species and on various membranes for the construction of BOD biosensor. The most commonly used membranes were polyvinyl alcohol, porous hydrophilic membranes etc.

A defined microbial consortium can be formed by conducting a systematic study, i.e. pre-testing of selected micro-organisms for use as a seeding material in BOD analysis of a wide variety of industrial effluents. Such a formulated consortium can be immobilized on suitable membrane, i.e. charged nylon membrane useful for BOD estimation. Suitability of charges nylon membrane lies in the specific binding between negatively charged bacterial cell and positively charged nylon membrane. So the advantages of the nylon membrane over the other membranes are : The dual binding, i.e. Adsorption as well as entrapment, thus resulting in a more stable immobilized membrane. Such specific Microbial consortium based BOD analytical devices, may find great application in monitoring of the degree of pollutional strength, in a wide variety of Industrial waste water within a very short time.

Biosensors can be used to indirectly measure BOD via a fast (usually <30 min) to be determined BOD substitute and a corresponding calibration curve method (pioneered by Karube et al., 1977). Consequently, biosensors are now commercially available, but they do have several limitations such as their high maintenance costs, limited run lengths due to the need for reactivation, and the inability to respond to changing quality characteristics as would normally occur in wastewater treatment streams; e.g. diffusion processes of the biodegradable organic matter into the membrane and different responses by different microbial species which lead to problems with the reproducibility of result (Praet et al., 1995). Another important limitation is the uncertainty associated with the calibration function for translating the BOD substitute into the real BOD.

Fluorescent RedOx Indicator

A surrogate to BOD_5 has been developed using a resazurin derivative which reveals the extent of oxygen uptake by micro-organisms for organic matter mineralization. A cross-validation performed on 109 samples in Europe and the United-States showed a strict statistical equivalence between results from both methods. The French start-up Envolure (Montpellier, France) offers the kit ENVERDI which enables the users to perform up to 40 BOD_5 simultaneously in 48 hours in a single 96-wells microplate.

Software Sensor

Rustum et al. (2008) proposed the use of the KSOM to develop intelligent models for making rapid inferences about BOD using other easy to measure water quality parameters, which, unlike BOD, can be obtained directly and reliably using on-line hardware sensors. This will make the use of BOD for on-line process monitoring and control a more plausible proposition. In comparison to other data-driven modeling paradigms such as multi-layer perceptrons artificial neural networks (MLP ANN) and classical multi-variate regression analysis, the KSOM is not negatively affected by missing data. Moreover, time sequencing of data is not a problem when compared to classical time series analysis.

Dissolved Oxygen Probes: Membrane and luminescence

Since the publication of a simple, accurate and direct dissolved oxygen analytical procedure by Winkler, the analysis of dissolved oxygen levels for water has been key to the determination of surface water purity and ecological wellness. The Winkler method is still one of only two analytical techniques used to calibrate oxygen electrode meters; the other procedure is based on oxygen solubility at saturation as per Henry's law. Though many researchers have refined the Winkler analysis to dissolved oxygen levels in the low PPB range, the method does not lend itself to automation.

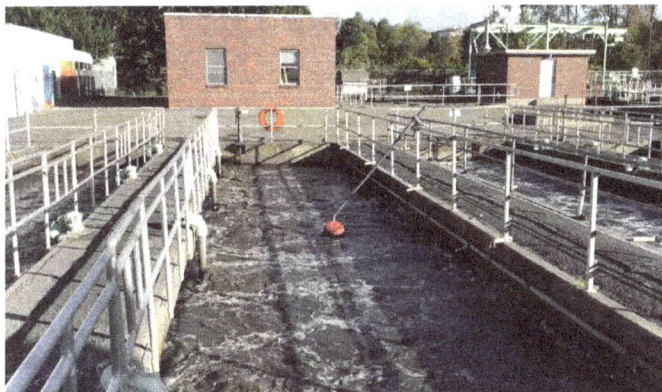

Dissolved oxygen sensor in a sewage treatment plant used as a feedback loop
to control the blowers in an aeration system.

The development of an analytical instrument that utilizes the reduction-oxidation (redox) chemistry of oxygen in the presence of dissimilar metal electrodes was introduced during the 1950s. This redox electrode (also known as dissolved oxygen sensor) utilized an oxygen-permeable membrane to allow the diffusion of the gas into an electrochemical cell and its concentration determined by polarographic or galvanic electrodes. This analytical method is sensitive and accurate to down to levels of ± 0.1 mg/l dissolved oxygen. Calibration of the redox electrode of this membrane electrode still requires the use of the Henry's law table or the Winkler test for dissolved oxygen.

During the last two decades, a new form of electrode was developed based on the luminescence emission of a photo active chemical compound and the quenching of that emission by oxygen. It is also called optical dissolved oxygen sensor. This quenching photophysics mechanism is described by the Stern–Volmer equation for dissolved oxygen in a solution:

$$I_0 / I = 1 + K_{SV} [O_2]$$

- I : Luminescence in the presence of oxygen

- I_0 : Luminescence in the absence of oxygen

- K_{SV} : Stern-Volmer constant for oxygen quenching

- $[O_2]$: Dissolved oxygen concentration

The determination of oxygen concentration by luminescence quenching has a linear response over a broad range of oxygen concentrations and has excellent accuracy and reproducibility. There are several recognized EPA methods for the measurement of dissolved oxygen for BOD, including the following methods:

- Standard Methods for the Examination of Water and Wastewater, Method 4500 O.

- In-Situ Inc. Method 1003-8-2009 Biochemical Oxygen Demand (BOD) Measurement by Optical Probe.

Test Limitations

The test method involves variables limiting reproducibility. Tests normally show observations varying plus or minus ten to twenty percent around the mean.

Toxicity

Some wastes contain chemicals capable of suppressing microbiological growth or activity. Potential sources include industrial wastes, antibiotics in pharmaceutical or medical wastes, sanitizers in food processing or commercial cleaning facilities, chlorination disinfection used following conventional sewage treatment, and odor-control formulations used in sanitary waste holding tanks in passenger vehicles or portable toilets. Suppression of the microbial community oxidizing the waste will lower the test result.

Appropriate Microbial Population

The test relies upon a microbial ecosystem with enzymes capable of oxidizing the available organic material. Some waste waters, such as those from biological secondary sewage treatment, will already contain a large population of microorganisms acclimated to the water being tested. An appreciable portion of the waste may be utilized during the holding period prior to commencement of the test procedure. On the other hand, organic wastes from industrial sources may require specialized enzymes. Microbial populations from standard seed sources may take some time to produce those enzymes. A specialized seed culture may be appropriate to reflect conditions of an evolved ecosystem in the receiving waters.

Carbonaceous Biochemical Oxygen Demand

Carbonaceous BOD (cBOD), and its cousins Biochemical Oxygen Demand (BOD) and Chemical Oxygen Demand (COD), are essentially measurements of energy. They are the energy contained in the molecular bonds of the carbonaceous and nitrogenous organic substances in a wastewater stream.

Organic material is food for microbes; these microbes break the bonds and consume energy released in the process. In an aerobic environment, the breaking of bonds consumes oxygen. Thus, the profile of the types and quantities of molecular bonds in the molecules determine its "oxygen demand." The classic BOD_5 test is one way to estimate the potential of waste to consume oxygen. It involves seeding a sample with live microbes, incubating and waiting for five days, then reading the oxygen depletion to gauge the result. This method is time-tested, being in use for over 100 years, but it's messy, imprecise, and most importantly, slow. The time delay from sample to results means that test results are useless for detecting transient problems or engaging in real-time process control. Fortunately, ZAPS Technologies has an alternative to this slow process.

Hybrid Multispectral Analysis (HMA) provides a rapid and precise technique for characterizing the driving force behind oxygen demand in a real-time basis. In the HMA approach, high intensity light is used to identify molecular bonds and particles in the sample stream characterizing the potential demand of the oxidants and oxidizers present. The HMA method's use of light allows for a very rapid and precise characterization of the sample stream without the need of surrogate microbial seeds or other sample disruptions. The end result is a continuous, direct, and efficient form of monitoring.

The CBOD5 Test

The CBOD tests have the widest application in measuring waste loadings to treatment plants and in evaluating the CBOD-removal efficiency of such treatment systems. The test measures the molecular oxygen utilized during a specified incubation period for the biochemical degradation of organic material (carbonaceous demand) and the oxygen used to oxidize inorganic material such as sulfides and ferrous iron. It also may measure the amount of oxygen used to oxidize reduced forms of nitrogen (nitrogenous demand) unless their oxidation is prevented by an inhibitor. The seeding and dilution procedures provide an estimate of the CBOD at pH 6.5 to 7.5.

There are two recognized EPA methods for the measurement of CBOD:

- Standard Methods for the Examination of Water and Wastewater, Method 5210B.

- In-Situ Inc. Method 1004-8-2009 Carbonaceous Biochemical Oxygen Demand (CBOD) Measurement by Optical Probe.

Method Summary

Bring the sample to ambient room temperature. If pH of sample is <6.5 or >7.5 neutralize the sample to approximately a pH of 7.0 using either sulfuric acid or sodium hydroxide. Aliquots of the neutralized sample are transferred to 300 mL CBOD bottles. These CBOD samples must be at concentrations that will deplete by at least 2 mg/L dissolved oxygen (DO) and have at least 1 mg/L DO left after five days of incubation. Therefore, make enough dilutions (minimum of 3) of the prepared sample to bracket the predicted CBOD.

The minimum aliquot volume transferred to a 300 mL CBOD bottle will be 3 mL as set by Standard Methods. If a smaller volume is needed to meet the DO depletion requirements, then you must make dilutions to the sample. Add approximately 0.1 g of Nitrification Inhibitor (2-chloro-6-(tri-chloro-methyl) pyridine) to each 300mL CBOD bottle before adding CBOD dilution water. If the sample is being prepared as a seeded sample, add enough prepared seed to the sample to achieve acceptable dissolved oxygen depletion. Add CBOD Dilution water to each CBOD sample bottle so as to completely fill the bottle with no air spaces or bubbles when the stopper is placed in the bottle.

Place the dissolved oxygen probe in the bottle and allow the dissolved oxygen meter to come to equilibrium. Allow the meter to come to equilibrium prior to accepting dissolved oxygen value. Record the DO of the sample, stopper the bottle, add DI water to the water seal if needed, cap the water seal, and incubate for 5 days at 20 °C ± 1 °C. Exclude light to avoid growth of algae in the bottles during incubation.

Upon completion of the 5-day incubation± 6 hours, record the DO of the depleted samples with a calibrated DO meter. Allow the meter to come to equilibrium prior to accepting dissolved oxygen value. Calculate the CBODs from the formula below. Only bottles, including seed controls, giving a minimum DO depletion of 2.0 mg/L and a residual DO of at least 1.0 mg/L after 5 days of incubation are considered to produce valid data, because at least 2.0 mg oxygen uptake per L is required to give a meaningful measure of oxygen uptake and at least 1.0 mg/L must remain throughout the test to ensure that insufficient DO does not affect the rate of oxidation of waste constituents.

Bacterial Seed CBOD Correction

Seed CBOD Uptake: Typically a 10, 20, and 30 mL sample of seed added to 3 separate CBOD bottles with approximately 0.1 g Nitrification Inhibitor and diluted with CBOD dilution water. Run these QC samples with each batch of seeded CBOD. Calculate the DO uptake per mL of seed added to each bottle using either the slope method or the ratio method.

For the slope method, plot DO depletion in milligrams per liter versus mLs of seed for all seed control bottles having a 2.0 mg/L depletion and 1.0 minimum residual DO. The plot should present a straight line for which the slope indicates DO depletion per mL of seed. The DO-axis intercept is oxygen depletion caused by the dilution water and should be less than 0.20 mg/L.

For the ratio method, divide the DO depletion by the volume of seed in mLs for each seed control bottle having a 2.0 mg/L depletion and greater than 1.0 mg/L minimum residual DO and average the results.

CBOD Seed

The CBOD test is method defined. Factors such as bacterial seed viability, anoxic stress during the 5 days, and nitrogenous inhibition efficacy will produce method variability between duplicates, analysts and laboratories. Clear quality assurance and quality control limits must be developed to produce valid results.

Sample Toxicity

Wastewater by definition may contain pollutants that inhibit bacterial seed metabolisms or are toxic to the seed. In these cases, all samples should be seeded with a known amount of viable bacteria for the MBOD analysis. Toxicity or inhibition is observed in CBOD analysis when the calculated CBOD increases with progressive dilutions of the sample.

Appropriate Microbial Population

Selection of a viable microbial population for the CBOD analysis is key in obtaining valid results. The bacterial population needs both carbonaceous and nitrogenous strains present. Sources of viable bacterial seed can be primary clarifier effluent, non-disinfected secondary clarifier effluent or a commercial seed preparation. Each source should have clear quality assurance and quality control requirements set by the glucose-glutamic acid check sample.

Glucose-glutamic Acid Check Sample

Transfer a known amount of glucose-glutamic acid solution to a CBOD bottle and add sufficient seed to achieve acceptable dissolved oxygen depletion. Fill CBOD bottle with CBOD dilution water and nitrification inhibitor. Determine the 5 Day CBOD. Passing results will have a CBOD of 198 (+ 30.5) mg/L. Run these check samples with each batch of CBOD samples. It is important to realize that glucose-glutamic acid is not intended to be an accuracy check in the test. Its sole purpose is to demonstrate that the seed is viable and metabolizing in the proper range of activity under the conditions of the test.

Regulatory use

In order to reduce a wastewater plants BOD_5 values to meet regulatory compliance requirements, some plant operators try to suppress nitrification when they are not required to meet ammonia limits. This practice usually results in increased effluent toxicity and oxygen demand on the receiving waters. Therefore, to eliminate this situation and because the BOD_5 test is not reflective of effluent quality under nitrifying conditions, the wastewater plant should:

1. Perform parallel $CBOD_5$ and BOD_5 tests to indicate whether there is a problem with BOD_5 compliance due to nitrification in the BOD_5 test results and that the $CBOD_5$ is not directly correlated with the BOD_5 test results, and

2. Baseline wastewater plant influent and effluent ammonia, nitrite and nitrate data (same frequency and duration as the parallel $CBOD_5$ and BOD_5 data) have been provided to perform mass balances for nitrification inhibition.

The results of these analysis can show that $CBOD_5$ should be utilized for regulatory compliance with wastewater discharge requirements.

BOD technically stands for Biological Oxygen Demand

BOD_5 means the test has been run for 5 days.

C-BOD means only the Carbonaceous Biochemical Oxygen Demand

N-BOD means Nitrogenous Biochemical Oxygen Demand

BOD_5 typically includes C-BOD and N-BOD unless one or the other is inhibited.

What are the main differences of C-BOD vs. BOD_5?

There are two completely different tests-a C-BOD test and a BOD5 test. Many times a C-BOD vs. BOD5 test is needed due to conditions at a plant. In some places where the nitrification of ammonia may not be complete (i.e., incomplete conversion of ammonia (NH_3) to nitrate (NO_3)) or where too high levels of amines or ammonia are present, false BOD readings may occur. This can occur in municipal lagoons, chemical plants or refineries. For lagoon (pond) treatment systems or other situations where this may occur, it is recommended that a Carbonaceous Biochemical Oxygen Demand (CBOD or Inhibited BOD) should be reported and used in place of 5-day Biochemical Oxygen Demand (BOD_5). A special chemical is added to kill the autotrophic bacteria so Nitrification is inhibited so that only the oxidation of COD occurs.

The biochemical oxygen demand (BOD) test tries to closely model an aerobic wastewater treatment system and the natural aquatic ecosystem. It measures oxygen taken up by the bacteria during the oxidation of organic matter. The test usually runs for a five-day period, but can run 7 or 10 days as well, depending on specific sample circumstances.

This conversion of carbon to cells is the synthesis reaction requiring about 0.5 to 0.6 lb O2/lbBOD (KgO2/Kg BOD). If the process is continued, a second oxygen demand is exerted for oxidizing the cells or digesting (stabilizing) the cells. This second phase is endogenous respiration, and requires an additional 0.8 to 0.9 lb O2/lb BOD (Kg/Kg). Pounds (kg) total oxygen required for carbonaceous BOD removal can range from 0.7lb/lb (kg/kg) BOD for high rate activated sludge with short detention of Biomass (low sludge age) up to 1.5 lb/lb (kg/kg) BOD for extended aeration with long detention of Biomass (MLSS, i.e., (long sludge age) in the bio system.

N-BOD means Nitrogenous Biochemical Oxygen Demand - All forms of 'reactive nitrogen' in urine and proteins (urea, uric acids, ammonia, amino acids, nitrates) are nutrients for algae and aquatic plant growth.

The nitrogenous waste in municipal and industrial sewage is used by autotrophic bacteria and they use a significant amount of oxygen as an energy source and convert ammonia to nitrates.

This phenomenon is called N-BOD or Nitrogenous Biochemical Oxygen Demand. The nutrient enrichment 'pollution' contributes to the eutrophication of lakes, rivers and water bodies when discharged in a final effluent.

The TKN (Total Kjeldahl Nitrogen) test measures the amount of reactive nitrogen (ammonia and organic nitrogen) in the sample that can be used by autotrophic bacteria and when they do, require oxygen, thus exerting a N-BOD, which would be equal to 4.6 x TKN mg/l.

Theoretically you can calculate Total Biological Oxygen Demand of any influent = 1.5 x BOD_5 + 4.6 x TKN.

While Carbonaceous BOD theoretically should require ~1.5 parts of O2 per part of BOD to be removed, nitrogenous BOD is significantly higher.

For nitrogenous BOD the demand for oxygen is 4.6 lb O2/ lb BOD (4.6Kg/Kg) removed. To achieve nitrogenous conversion of ammonia to nitrate requires longer aeration time with low food to microorganism ratio, i.e., much sludge MLSS (M) with low food supply (F). This condition results in a long sludge age, which promotes nitrification.

Chemical Oxygen Demand

The chemical oxygen demand (COD) is a measure of water and wastewater quality. The COD test is often used to monitor water treatment plant efficiency. This test is based on the fact that a strong oxidizing agent, under acidic conditions, can fully oxidize almost any organic compound to carbon dioxide. The COD is the amount of oxygen consumed to chemically oxidize organic water contaminants to inorganic end products.

The COD is often measured using a strong oxidant (e.g. potassium dichromate, potassium iodate, potassium permanganate) under acidic conditions. A known excess amount of the oxidant is added to the sample. Once oxidation is complete, the concentration of organics in the sample is calculated by measuring the amount of oxidant remaining in the solution. This is usually done by titration, using an indicator solution. COD is expressed in mg/L, which indicates the mass of oxygen consumed per liter of solution.

The COD test only requires 2-3 hours, while the Biochemical (or Biological) Oxygen Demand (BOD) test requires 5 days. It measures all organic contaminants, including those that are not biodegradable. There is a relationship between BOD and COD for each specific sample, but it must be established empirically. COD test results can then be used to estimate the BOD of a given sample. Unlike for the BOD test, toxic compounds (such as heavy metals and cyanides) in the samples to be analyzed do not have an effect on the oxidants used in the COD test. Therefore, the COD test can be used to measure the strength of wastes that are too toxic for the BOD test. Some organic molecules (e.g., benzene, pyridine) are relatively resistant to dichromate oxidation and may give a falsely low COD.

Chemical oxygen demand is related to biochemical oxygen demand (BOD), another standard test for assaying the oxygen-demanding strength of waste waters. However, biochemicaloxygen demand only measures the amount of oxygen consumed by microbial oxidation and is most relevant to waters rich in organic matter. It is important to understand that COD and BOD do not necessarily measure the same types of oxygen consumption. For example, COD does not measure the oxygen-consuming potential associated with certain dissolved organic compounds such as acetate. However, acetate can be metabolized by microorganisms and would therefore be detected in an assay of BOD. In contrast, the oxygen-consuming potential of cellulose is not measured during a short-term BOD assay, but it is measured during a COD test.

The basis for the COD test is that nearly all organic compounds can be fully oxidized to carbon dioxide with a strong oxidizing agent under acidic conditions. The amount of oxygen required to oxidize an organic compound to carbon dioxide, ammonia, and water is given by:

$$C_nH_aO_bN_c + \left(n + \frac{a}{4} - \frac{b}{2} - \frac{3}{4}c\right)O_2 \rightarrow nCO_2 + \left(\frac{a}{2} - \frac{3}{2}c\right)H_2O + cNH_3$$

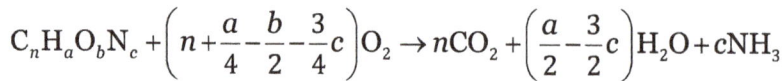

This expression does not include the oxygen demand caused by nitrification, the oxidation of ammonia into nitrate:

$$NH_3 + 2O_2 \rightarrow NO_3^- + H_3O^+$$

Dichromate, the oxidizing agent for COD determination, does not oxidize ammonia into nitrate, so nitrification is not included in the standard COD test.

The International Organization for Standardization describes a standard method for measuring chemical oxygen demand in ISO 6060.

Using Potassium Dichromate

Potassium dichromate is a strong oxidizing agent under acidic conditions. Acidity is usually achieved by the addition of sulfuric acid. The reaction of potassium dichromate with organic compounds is given by:

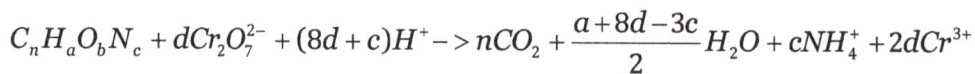

$$C_nH_aO_bN_c + dCr_2O_7^{2-} + (8d + c)H^+ -> nCO_2 + \frac{a + 8d - 3c}{2}H_2O + cNH_4^+ + 2dCr^{3+}$$

where $d = 2n/3 + a/6 - b/3 - c/2.$. Most commonly, a 0.25 N solution of potassium dichromate is used for COD determination, although for samples with COD below 50 mg/L, a lower concentration of potassium dichromate is preferred.

In the process of oxidizing the organic substances found in the water sample, potassium dichromate is reduced (since in all redox reactions, one reagent is oxidized and the other is reduced), forming Cr^{3+}. The amount of Cr^{3+} is determined after oxidization is complete, and is used as an indirect measure of the organic contents of the water sample.

Measurement of Excess

For all organic matter to be completely oxidized, an excess amount of potassium dichromate (or any oxidizing agent) must be present. Once oxidation is complete, the amount of excess potassium dichromate must be measured to ensure that the amount of Cr^{3+} can be determined with accuracy. To do so, the excess potassium dichromate is titrated with ferrous ammonium sulfate (FAS) until all of the excess oxidizing agent has been reduced to Cr^{3+}. Typically, the oxidation-reduction indicator ferroin is added during this titration step as well. Once all the excess dichromate has been reduced, the ferroin indicator changes from blue-green to a reddish brown. The amount of ferrous ammonium sulfate added is equivalent to the amount of excess potassium dichromate added to the original sample. Note: Ferroin indicator is bright red from commercially prepared sources, but when added to a digested sample containing potassium dichromate it exhibits a green hue. During the titration the color of the indicator changes from a green hue to a bright blue hue to a reddish brown upon reaching the endpoint. Ferroin indicator changes from red to pale blue when oxidized.

Preparation of Ferroin Indicator Reagent

A solution of 1.485 g 1,10-phenanthroline monohydrate is added to a solution of 695 mg $FeSO_4 \cdot 7H_2O$ in distilled water, and the resulting red solution is diluted to 100 mL.

Calculations

The following formula is used to calculate COD:

$$COD = \frac{8000(b-s)n}{\text{sample volume}}$$

where b is the volume of FAS used in the blank sample, s is the volume of FAS in the original sample, and n is the normality of FAS. If milliliters are used consistently for volume measurements, the result of the COD calculation is given in mg/L.

The COD can also be estimated from the concentration of oxidizable compound in the sample, based on its stoichiometric reaction with oxygen to yield CO_2 (assume all C goes to CO_2), H_2O (assume all H goes to H_2O), and NH_3 (assume all N goes to NH_3), using the following formula:

$$COD = (C/FW) \cdot (RMO) \cdot 32$$

Where

    C = Concentration of oxidizable compound in the sample,

    FW = Formula weight of the oxidizable compound in the sample,

    RMO = Ratio of the # of moles of oxygen to # of moles of oxidizable compound in their reaction to CO_2, water, and ammonia

For example, if a sample has 500 Wppm (Weight Parts per Million) of phenol:

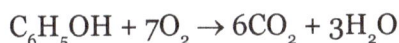

$$C_6H_5OH + 7O_2 \rightarrow 6CO_2 + 3H_2O$$

$$COD = (500/94) \cdot 7 \cdot 32 = 1191 \text{ Wppm}$$

Inorganic Interference

Some samples of water contain high levels of oxidizable inorganic materials which may interfere with the determination of COD. Because of its high concentration in most wastewater, chloride is often the most serious source of interference. Its reaction with potassium dichromate follows the equation:

$$6Cl^- + Cr_2O_7^{2-} + 14H^+ \rightarrow 3Cl_2 + 2Cr^{3+} + 7H_2O$$

Prior to the addition of other reagents, mercuric sulfate can be added to the sample to eliminate chloride interference.

The following table lists a number of other inorganic substances that may cause interference. The table also lists chemicals that may be used to eliminate such interference, and the compounds formed when the inorganic molecule is eliminated.

Inorganic molecule	Eliminated by	Elimination forms
Chloride	Mercuric sulfate	Mercuric chloride complex
Nitrite	Sulfamic acid	N_2 gas
Ferrous iron	-	-
Sulfides	-	-

Government Regulation

Many governments impose strict regulations regarding the maximum chemical oxygen demand allowed in waste water before they can be returned to the environment. For example, in Switzerland, a maximum oxygen demand between 200 and 1000 mg/L must be reached before waste water or industrial water can be returned to the environment.

Adsorbable Organic Halides

Adsorbable organic halides (AOX) is an organic sum parameter comprising such organics that contain chlorine, bromine or iodine atoms and are adsorbable to activated carbon. For AOX determination a particular volume of the wastewater sample is agitated sufficiently long with powdered activated carbon. Subsequently the activated carbon is separated by filtration using a membrane filter which retains the activated carbon (adsorption can also be executed in small activated carbon columns which are treated - after adsorption has been completed - in the same way as the loaded activated carbon removed by filtration). Then the membrane filter is incinerated together with the activated carbon in a stream of pure oxygen at temperatures around 900°C. The halogen atoms originally bound in organics adsorbed to the activated carbon form HCl, HBr, or HI, resp., which are contained in the exhaust gas of the incineration furnace and can be absorbed e.g. in acetic acid. Microcoulometric titration, an electrochemical quantification method, analyses chloride, bromide, or iodide, resp., of these acids. Bromide and iodide are calculated as chloride equivalents (one mol bromide or iodide is looked at as one mol chloride and is calculated as chloride mass), and the final chloride mass determined is related to the volume of the wastewater sample which had been subdued to activated carbon adsorption. The result is mg AOX (chloride)/l wastewater.

In the AOX analysis procedure, artefacts can easily be produced: First, also inorganic chloride adsorbs to a certain amount to activated carbon. This adsorbed inorganic chloride will also be detected e.g. by microcoulometric analysis of the incineration off-gass and may result in the so-called "chloride error". Secondly, in wastewaters with high TOC mainly represented by non-halogenated organic compounds a competition of halogenated and non-halogenated organic compounds for adsorption sites on the activated carbon occurs leading to a very low extent of halogenated organic molecules being adsorbed. This can be prevented by dilution of the wastewater sample. However, by dilution also the AOX is diluted which is disadvantageous if the AOX content of the sample is decreased to be below the detection limit of the method. AOX analyses must be performed in laboratory rooms where no halogenated organic solvents are used at all, because these volatiles would also adsorb on the activated carbon during the AOX procedure. In recent years, AOX analyses in the Institute of Wastewater Management of Hamburg University of Technology had been performed in a laboratory where a thermostatized chamber was located. When there was a leakage in the cooling system of the chamber, some fluorochlorohydrocarbons were volatilized in the laboratory leading to severe analytical errors in AOX determinations.

Other parts of organics contained in wastewaters (usually comprised in TOC or COD) are the organic sum parameters hydrocarbons, phenols, anionic surfactants, neutral surfactants, cationic surfactants etc. Methods for analyzing these organic sum parameters are also given in the "Standard Methods".

Determination of AOX

Persistent organic pollutants such as dichlorodiphenyltrichloroethane (DDT), polychlorinated biphenols, dioxins, are all assessed in AOX analysis. Generally, the higher the amount of chlorine in an organic compound, the more toxic it is considered. While there are several biochemical or electrochemical methods to remove organic halides, AOX has been preferred due to its low cost of operation and simplicity of design.

In a lab, the determination of AOX parameter consists of adsorption of organic halides from the sample on to an activated carbon. The activated carbon can be powdered or granular and adsorbed using microcolumns or a batch process, if the samples are rich in humic acids. Vigorous shaking is often employed in the event of a batch process to favor the adsorption of organic halide on to the activated carbon due to its electronegativity and presence of lone pairs. The inorganic halides that are also adsorbed are washed away using a strong acid such as nitric acid. The carbon with adsorbed organic halide is obtained by filtration, after which the filter containing the carbon is burnt in the presence of oxygen. While combustion of hydrocarbon part of the compounds form CO_2 and H_2O, halo acids are formed from the halogens. These haloacids are absorbed into acetic acid. Subsequent use of microcolumetric titration, an electrochemical quantification method, provides the AOX content in the sample. Using the dilution ratio, the total AOX content at the location can be estimated. Alternatively, the chlorinated compounds in the sample can be determined by using pentane extraction followed by capillary gas chromatography and electron capture (GC-ECD). The organic carbon that was remaining after the nitric acid purge can be analyzed using UV-persulfate wet oxidation followed by Infrared-detection (IR). Several other analytical techniques such as high performance liquid chromatography (HPLC) could also be implemented to quantify AOX levels. The general adsorption procedure is given below:

$$C^*_{(S)} + R - X(a.q) \rightarrow C^* - X - R_{(S)} + H_2O$$

Where $C^*_{(S)}$ is the activated carbon and $R - X$ is any organic halide.

$C^* - X - R_{(S)}$ is the organic halide - activated carbon complex that can be filtered out.

Treatment

Physical Separation

In water treatment plants, organic halides are adsorbed using GAC or PAC in agitated tanks. The loaded carbon is separated using a membrane made out of materials like polypropylene or cellulose nitrate. Measuring the AOX levels into and out of the treatment zone shows a drop in organic halide concentrations. Some processes use a two-step GAC filtration to remove AOX precursors, and thus reduce the amount of AOX in treated waters. A two step filtration process consists of two GAC filters in series. The first filter is loaded with exhausted GAC, while the second filter is loaded with fresh GAC. This set up is preferred for its increased efficiency and higher throughput capacity. The GAC is replaced cyclically and the extracted organic halide-carbon mixture is then sent for subsequent biological or chemical treatment such as ozonation to regenerate the GAC. Often, these chemical treatments, while effective, pose economical challenges to the treatment plants.

Biological Treatment

A more economically attractive option for treatment of the organic halides is through utilization of biological agents. Recently, bacteria *(Ancylobacter aquaticus)*, fungi *(Phanerochaete chrysosporium* and *Coiriolus versicolor)*, or synthetic enzymes have been used in the degradation of chlorinated organic compounds. The microorganisms degrade halocompounds using either aerobic or anaerobic processes. The mechanisms of degradation include utilization of the compound as carbon source for energy, cometabolite, or as an electron acceptor. Note that enzymatic or microbial action could be regulated through feedback inhibition-the final product in the series inhibits a reaction in the process. An example of a microbe that can degrade AOX is shown below in Figures.

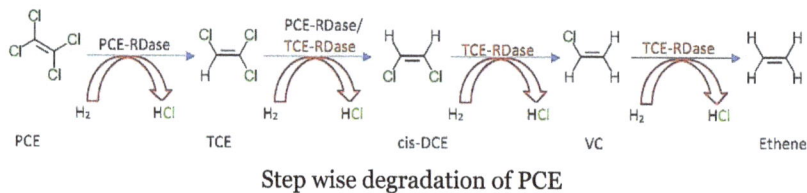

Step wise degradation of PCE

A sample dechlorination of chlorinated aliphatic hydrocarbons (CAHs) such as perchloroethylene (PCE) by *Dehalococcoides ethenogenes* has been illustrated above. PCE is one of the highly chlorinated CAHs with no known microorganisms capable of aerobic degradation. The high electronegative character of PCE renders oxidizing agent capabilities through accepting electrons by co-metabolism or dehalorespiration. In a co-metabolism, the reduction of PCE is made feasible by the utilization of a primary metabolite for carbon and energy source. In dehalorespiration, the electron transfer from oxidation of small molecules (H_2 is the major source; but, glucose, acetate, formate, and methanol can also be used) to PCE generates energy required for the bacterial growth. The hydrogen involved in this mechanism is often a product of another process such as fermentation of simple molecules like sugars or other complex molecules like fatty acids. Moreover, due to competition from methanogens for H_2, low H_2 concentrations are favored by dechlorinating bacteria, and is often established through slow-release fermentation compounds such as fatty acids and decaying bacterial biomass. While several enzymes and electron carriers are involved in process, two enzymes perform the dechlorination reactions–PCE reductive dehydrogenase (PCE-RDase) and TCE reductive dehydrogenase (TCE-RDase). The PCE-RDase is normally found freely in cytoplasm while the TCE-RDase is found attached to the exterior cytoplasmic membrane. These enzymes normally utilize a metal ion cluster like Fe-S cluster to complete electron transfer cycle. Hydrogen is oxidized to generate two protons and two electrons. The removal of first chloride, which is performed by PCE-RDase, reduces PCE into TCE by reductive dehalogenation, where a hydride replaces the chlorine. The chloride lost from PCE gains the two electrons and the proton that accompanies them to form HCl. TCE can be reduced to *cis*-dichloroethene (cis-DCE) by either PCE-RDase or TCE-RDase. Subsequent reductions to vinyl chloride (VC) and ethylene are performed by TCE-RDase. The dechlorination of PCE to cis-DCE is faster and thermodynamically more favorable than dechlorination of cis-DCE to VC. The transformation of VC to ethylene is the slowest step of the process and hence limits the overall rate of the reaction. The rate of reductive dechlorination is also directly correlated with the number of chlorine atoms, and as such, it decreases with a decreasing number of chlorine atoms. In addition, while several groups of bacteria such as *Desulfomonile, Dehalobacter, Desulfuromonas*...etc. can perform the dehalogenation of PCE to TCE, only the *Dehalococcoides* group can perform the complete reductive dechlorination from PCE to ethene.

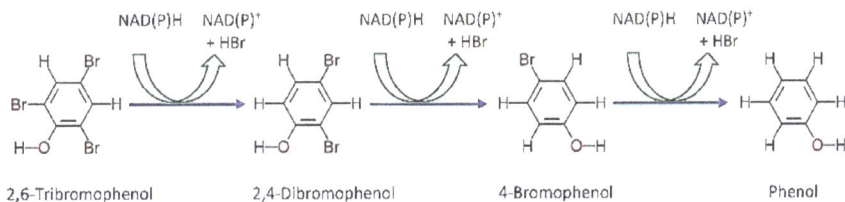

2,4,6-TBP degradation by *Ochrobactrum sp.*

In addition to dechlorination of CAHs, microbes have also been reported to act on chlorinated aromatic hydrocarbons. An example of a reaction where aromatic AOX content has been reduced is demonstrated in figure above. While little is known about the dehalogenation mechanisms of polyhalogenated phenols (PHPs) and polyhalogenated benzenes (PHBs), regioselectivity for halide location on the aromatic ring was observed. This regioselectivity is however dominated by both redox potentials for the reaction and the microbe's familiarity to the reaction. Moreover, due to the specificity of most microbes along with complex aromatic structures, in order to achieve a complete dehalogenation, a mixture of more than one species of bacteria and/or fungi (often known as a consortium) is utilized. The reaction in figure 2 shows the reductive debromination of 2,4,6-tribromophenol (2,4,6-TBP) by *Ochrabactrum*. Based on the relative degradation of the molecule along with analytical results, it has been postulated that degradation of 2,4,6-TBP proceeds through debromination of *ortho*-bromine in the first step by a dehalogenase to yield 2,4-dibromophenol (2,4-DBP). Since there are two *ortho* bromines, debromination of either *ortho* carbons would yield the same product . Other species such as *Pseudomonas galthei* or *Azotobacter sp.* showed preference for *para*-halide over the *meta*- or *ortho* -halides. For example, the *Azotobacter sp.* degrades 2,4,6-trichlorophenol (2,4,6-TCP) into 2,6-dichlorohydroquinone due to TCP-4-monooxygenase selectivity differences between *ortho*- and *para*-halide. These differences in regioselectivity between the species can be attributed to the specificity of the 3-dimensional enzyme structure and its hindrance from steric interactions. It has been postulated that a proton lost by the phenol group of 2,4,6-TBP resulting in the formation of a negatively charged halo-phenolate ion. Subsequent attack of the *para*-carbon with a hydride anion from NAD(P)H in a nucleophilic attack manner and resonance rearrangement results in substitution of bromine with hydride and formation of 2,4-DBP. Subsequent steps in a similar pattern yield 2-bromophenol, and phenol in the final step. Phenol can be metabolized by microorganisms to make methane and carbon dioxide or can be extracted easier than AOXs.

Total Dissolved Solids

Total Dissolved Solids, also known as TDS, are inorganic compounds that are found in water such as salts, heavy metals and some traces of organic compounds that are dissolved in water.

Excluding the organic matters that are sometimes naturally present in water and the environment, some of these compounds or substances can be essential in life. But, it can be harmful when taken more than the desired amount needed by the body.

The total dissolved solids present in water are one of the leading causes of turbidity and sediments in drinking water. When left unfiltered, total dissolved solids can be the cause various diseases.

A total dissolved solid (TDS) is a measure of the combined total of organic and inorganic substances contained in a liquid. This includes anything present in water other than the pure H_2O molecules. These solids are primarily minerals, salts and organic matter that can be a general indicator of water quality.

The TDS in Drinking Water

The TDS in drinking water comes from natural water sources, sewage, urban run-off, industrial wastewater and chemicals used in water treatment process, and the hardware or piping used to distribute water. In US, higher TDS was brought by natural environment features like salt deposits, mineral springs. sea water intrusion, and carbonate deposits. Other sources may include anti-skid materials, salts used for road de-icing, stormwater, and agricultural runoff, water treatment chemicals, and point/non-point wastewater discharges.

In general, the total dissolved solids concentration is the total cations (positively charged) and anions (negatively charged) ions in the water. Thus, the total dissolved solids test gives a qualitative measure of the amount of dissolved ions but does not tell us the nature or ion relationships. In addition, the test does not provide us insight into the specific water quality issues, such as Elevated Hardness (mineral content in water) , Salty Taste, or Corrosiveness (also called as aggressive water which is how water dissolves with other materials).

Therefore, the total dissolved solids test is used as an indicator test to determine the general quality of the water. The sources of total dissolved solids can include all of the dissolved cations and anions, but the following table can be used as a generalization of the relationship of TDS to water quality problems.

Cations combined with Carbonates $CaCO_3$, $MgCO_3$ etc	Associated with hardness, scale formation, bitter taste
Cations combined with Chloride NaCl, KCl	Salty or brackish taste, increase corrosivity

What composes total dissolved solids and how do they contaminate our water supply?

There are different substances that comprise the total dissolved solids in drinking water. As a natural flora of water and the environment, bacteria and viruses can be found in total dissolved solids, these are the organic compounds found in drinking water. Chemicals found in the water and water supply include heavy metals, salts, and pharmaceutical drugs that are caused by human waste materials which contaminate the water and water supply

Water that comes from springs, lakes and waterfalls has natural microorganisms and salts. This will in turn go to the public water treatment and are stored for supplying the community. Not only that but also phytoplankton, one of the natural floras of water, can sometimes be found. A phytoplankton is a type of microscopic plant that drifts off to different bodies of water.

Sometimes, chemicals such as iron, potassium, sodium and other chemicals that are known to man are present in drinking water too. These chemicals are caused by human waste products that contaminate these water sources. Not to mention the volatile organic compounds, they also contaminate the water by leaking through the water supply by soil.

Volatile organic compounds, also known as VOC's, are rapidly evaporating compounds that are chemically designed for specific use at home, at school and anywhere you can think of.

Other Reason Why Solids End Up Dissolved in Water

Mineral springs contain water with high levels of dissolved solids, because the water has flowed through a region where the rocks have a high salt content. For instance, the water in the Prairie provinces in US tends to have high levels of dissolved solids, because of high amounts of calcium and magnesium in the ground.

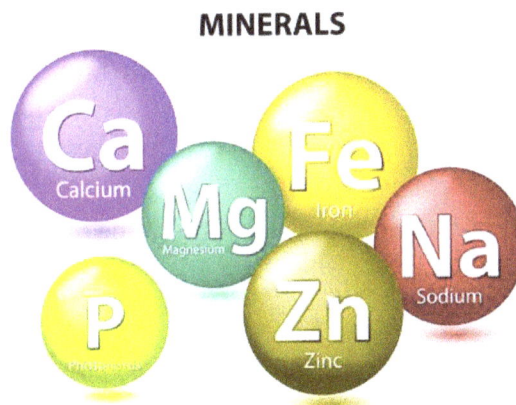

MINERALS

These minerals can also come from human activities. Agricultural and urban runoff can carry excess minerals into water sources, as can wastewater discharges, industrial wastewater and salt that is used to de-ice roads.

The Acceptable Total Dissolved Solids (TDS) Level in Drinking Water

Total Dissolved Solids (TDS) is measured in milligrams per unit volume of water (mg/L) and also referred to as parts per million (ppm). For drinking water, the maximum concentration level set by EPA is 500 mg/L.

Need for Measuring the TDS Levels in your Water

Numerous water supplies exceed this level. When TDS levels exceed 1000mg/L it is generally considered unfit for human consumption. A high level of TDS is an indicator of potential concerns, and appeals for further investigation.

Most often, high levels of TDS are caused by the presence of potassium, chlorides and sodium. These ions have little or no short-term effects, but toxic ions (lead arsenic, cadmium, nitrate and others) may also be dissolved in the water.

Even the best water purification systems on the market require monitoring for TDS to ensure the filters and/or membranes are effectively removing unwanted particles and bacteria from your water. Here are other applications of the importance of TDS level:

- Taste / Health

High TDS results in undesirable taste which could be salty, bitter, or metallic. It could also indicate the presence of toxic minerals. The EPA's recommended maximum level of TDS in water is 500mg/L (500ppm).

- Filter performance

Test your water to make sure the reverse osmosis or other type of water filter or water purification system has a high rejection rate and know when to change your filter (or membrane) cartridges.

- Hardness (and Water Softeners)

High TDS point out the Hard water, which causes scale build-up in pipes and valves. This eventually restricts performance.

- Aquariums / Aquaculture

A constant level of minerals is needed for aquatic life. The water in an aquarium or tank must have the same levels of TDS and pH as the fish and reef's original habitat.

- Hydroponics

TDS is the best measurement of the nutrient concentration in a hydroponic solution.

- Pools and Spas

TDS levels must be monitored to prevent maintenance problems

- Commercial / Industrial

High TDS levels could prevent the functions of certain applications, such as boilers and cooling towers, food and water production and etc.

- Colloidal silver water

TDS levels must be controlled prior to making colloidal silver.

- Coffee and Food Service

For a truly great cup of coffee, proper TDS levels must be maintained.

- Car Washing and Window Cleaning

It gives the best result when cleaning.

The Ways to Remove Total Dissolved Solids in Drinking Water

These chemicals are absorbed by the soils that eventually contaminate the water supplies. Though some of the total dissolved solids are natural floras, filtration and purification is an effective way to remove these chemicals out from our drinking water.

To give you a short process of water filtration and purification, the water coming from the water sources goes to the public water treatment to undergo filtering. In this process, large particles or substances are filtered out and removed from the water.

Sometimes, in this process, small particles can get through the filters. This is still causing turbidity in drinking water and is not yet safe to be consumed. Once it undergoes filtration, it is then purified using certain methods. Depending on the company or water treatment facility, some use chemicals, radiation or micro-filtration and chlorination to remove microscopic particles still present in water.

In the process of removing dissolved solids in drinking water with the use of chemicals as reagents, these reagents help to filter these substances out. Chemicals such as aluminum sulphate and liquid chlorine are used. These chemicals are flocculating chemicals that help in water filtration. What they do is that they clump the particles together until they form a larger particle.

These flocks can then be easily removed or filtered out. Some chemicals used are fluorosilicic acid, sodium silicofluoride and sodium fluoride. These chemicals give a unique reaction. These chemicals give off energy by creating heat, when these chemicals are combined with water they burn thus killing the bacteria, viruses, protozoan and other organic microorganism found in water.

Once the drinking water is filtered of the impurities, it still has a strange taste that is not potable yet. This is when calcium hydroxide was added to water. Calcium hydroxide, also known as lime water, is a self-regulating chemical that does not affect the alkalinity and the acidity of water. So, it makes it safe to be drink and will have the taste of natural water.

While some water treatment facility uses chemical reagent to filter and purify drinking, some public water treatment facility uses filtration, chlorination and Ultra Violet radiation to purify drinking water. The process of these treatment starts when water coming from water sources goes to the water supply for particles to be filtered out. Once filtered, it will remain in the water supply for chlorination.

When chlorine is added, it will remain in the water storage until the smell of chlorine cannot be detected. Since chlorine is a derivative of salt and sodium, sometimes it cannot be easily dissolved. When the chlorination process is done, it goes to a special process where radiation is involved.

With the help of Ultra Violet rays, water is then radiated to make sure that the remaining substances of chlorine will be dissolved and the microorganisms that made it through the filtration and chlorination process were killed. These residues turn into sediments found at the bottom of water and is the cause of turbidity in water.

Safely Purify Water

Well, after knowing a few of the processes to filter and purify water, different questions are forming deep in our head. Like, how safe is it to be consumed? How will it affect our health and our family as well?

Since we all know that radiation is harmful to our health and chemicals. When defined, it always leads to toxicity and disease. However, there is a way to purify water from these contaminants and remove the impurities safely and effectively without the use of chemicals or radiation.

 Another reliable way to purify water is boiling. This is the most basic way of purifying water. Also, when purifying water from doubtful water sources, you must not forget to filter the water as it may still have some sediment or solid particles.

In using a water filter, it is also a good way to remove the bacteria in water. Carbon filter gets rid of the awful tastes and chemicals while iodine-coated filter can remove bacteria and viruses. After treating the water, you can already drink the water and you may notice the water without any taste.

Now you can make sure that your family is drinking a safe and potable water without the undergoing these processes and without the presence of sediments or microorganisms that can cause disease.

The Berkey Water Filter aids in removing any kind of inorganic compounds, microorganisms and other contaminants. Using any water source, filtering it with the Berkey Water Filter will remove health threats to your water. It only retains the mineral that is why it retains the standard TSD level while leaving your water tastes better. With the help of Berkey you can now assure that your family will remain healthy, hydrated and free from harm.

Hydrological Simulation

Hydrologic transport models are used to mathematically analyze movement of TDS within river systems. The most common models address surface runoff, allowing variation in land use type, topography, soil type, vegetative cover, precipitation, and land management practice (e.g. the application rate of a fertilizer). Runoff models have evolved to a good degree of accuracy and permit the evaluation of alternative land management practices upon impacts to stream water quality.

Pyramid Lake, Nevada receives dissolved solids from the Truckee River.

Basin models are used to more comprehensively evaluate total dissolved solids within a catchment basin and dynamically along various stream reaches. The DSSAM model was developed by the U.S. Environmental Protection Agency (EPA). This hydrology transport model is actually based upon the pollutant-loading metric called "Total Maximum Daily Load" (TMDL), which addresses TDS and other specific chemical pollutants. The success of this model contributed to the Agency's broadened commitment to the use of the underlying TMDL protocol in its national policy for management of many river systems in the United States.

Practical Implications

Aquarium at Bristol Zoo, England. Maintenance of filters becomes costly with high TDS.

When measuring water treated with water softeners, high levels of total dissolved solids do not correlate to hard water, as water softeners do not reduce TDS; rather, they replace magnesium and calcium ions, which cause hard water, with an equal charge of sodium or potassium ions, e.g. $Ca^{2+} \rightleftharpoons 2$ Na^+, leaving overall TDS unchanged or even increased. Hard water can cause scale buildup in pipes, valves, and filters, reducing performance and adding to system maintenance costs. These effects can be seen in aquariums, spas, swimming pools, and reverse osmosis water treatment systems.

Typically, in these applications, total dissolved solids are tested frequently, and filtration membranes are checked in order to prevent adverse effects.

In the case of hydroponics and aquaculture, TDS is often monitored in order to create a water quality environment favorable for organism productivity. For freshwater oysters, trouts, and other high value seafood, highest productivity and economic returns are achieved by mimicking the TDS and pH levels of each species' native environment. For hydroponic uses, total dissolved solids is considered one of the best indices of nutrient availability for the aquatic plants being grown.

Because the threshold of acceptable aesthetic criteria for human drinking water is 500 mg/l, there is no general concern for odor, taste, and color at a level much lower than is required for harm. A number of studies have been conducted and indicate various species' reactions range from intolerance to outright toxicity due to elevated TDS. The numerical results must be interpreted cautiously, as true toxicity outcomes will relate to specific chemical constituents. Nevertheless, some numerical information is a useful guide to the nature of risks in exposing aquatic organisms or terrestrial animals to high TDS levels. Most aquatic ecosystems involving mixed fish fauna can tolerate TDS levels of 1000 mg/l.

The Fathead minnow (*Pimephales promelas*), for example, realizes an LD_{50} concentration of 5600 ppm based upon a 96-hour exposure. LD50 is the concentration required to produce a lethal effect on 50 percent of the exposed population. *Daphnia magna*, a good example of a primary member of the food chain, is a small planktonic crustacean, about 0.5 mm in length, having an LD50 of about 10,000 ppm TDS for a 96-hour exposure.

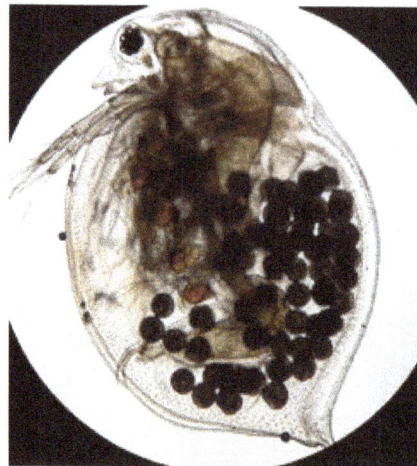

Daphnia magna with eggs

Spawning fishes and juveniles appear to be more sensitive to high TDS levels. For example, it was found that concentrations of 350 mg/l TDS reduced spawning of Striped bass (*Morone saxatilis*) in the San Francisco Bay-Delta region, and that concentrations below 200 mg/l promoted even healthier spawning conditions. In the Truckee River, EPA found that juvenile Lahontan cutthroat trout were subject to higher mortality when exposed to thermal pollution stress combined with high total dissolved solids concentrations.

For terrestrial animals, poultry typically possess a safe upper limit of TDS exposure of approximately 2900 mg/l, whereas dairy cattle are measured to have a safe upper limit of about 7100 mg/l. Research has shown that exposure to TDS is compounded in toxicity when other stressors are present, such as abnormal pH, high turbidity, or reduced dissolved oxygen with the latter stressor acting only in the case of animalia.

In countries with often-unsafe/unclean tap water supplies, such as much of India, the TDS of drinking water is often checked by technicians to gauge how effectively their RO/Water Filtration devices are working. While TDS readings will not give an answer as to the amount of microorganisms present in a sample of water, they can get a good idea as to the efficiency of the filter by how much TDS is present.

Total Suspended Solids

Total Suspended Solids (TSS) is the portion of fine particulate matter that remains in suspension in water. It measures a similar property to turbidity, but provides an actual weight of particulate matter for a given volume of sample (usually mg/l).

TSS are particles that are larger than 2 microns found in the water column. Anything smaller than this is called a dissolved solid. The majority of suspended solids are made up of inorganic materials, although bacteria and algae can contribute to total solid levels. These solids include anything floating through the water such as gravel, silt, sand or clay. Another contributor to TSS is the decomposing of plants and animals meaning small particles break away from the organism becoming a suspended solid in the water. TSS is significant in regards to the aesthetics of the water, as the more suspended solids that are present the less clear the water will become.

Some suspended solids can settle to the bottom of the body of water over time along with heavier particles such as sand and gravel. This is usually present in areas of water that are shallower due to slow water flow. This settling improves water clarity however the increased silt can smother eggs and benthic organism.

TSS is the most visible indicator of water quality. It is considered that clear water is usually considered healthy water. It is especially cause for concern if the water becomes murky in a previously clear body of water. Excessive suspended solids can be cause for concern for aquatic and human life as well as impede navigation and increase flooding risks.

First, let's consider some implications of total suspended solids (TSS).

-High concentrations of suspended solids may settle out onto a streambed or lake bottom and cover aquatic organisms, eggs, or macro-invertebrate larva. This coating can prevent sufficient oxygen transfer and result in the death of buried organisms.

-High concentrations of suspended solids decrease the effectiveness of drinking water disinfection agents by allowing microorganisms to "hide" from disinfectants within solid aggregates. This is one of the reasons the TSS, or turbidity, is removed in drinking water treatment facilities.

-Many organic and inorganic pollutants sorb to soils, so that the pollutant concentrations on the solids are high. Thus, sorbed pollutants (and solids) can be transported elsewhere in river and lake systems, resulting in the exposure of organisms to pollutants away from the point source.

Measurement

TSS of a water or wastewater sample is determined by pouring a carefully measured volume of water (typically one litre; but less if the particulate density is high, or as much as two or three litres for very clean water) through a pre-weighed filter of a specified pore size, then weighing the filter again after the drying process that removes all water on the filter. Filters for TSS measurements are typically composed of glass fibres. The gain in weight is a dry weight measure of the particulates present in the water sample expressed in units derived or calculated from the volume of water filtered (typically milligrams per litre or mg/L).

If the water contains an appreciable amount of dissolved substances (as certainly would be the case when measuring TSS in seawater), these will add to the weight of the filter as it is dried. Therefore it is necessary to "wash" the filter and sample with deionized water after filtering the sample and before drying the filter. Failure to add this step is a fairly common mistake made by inexperienced laboratory technicians working with sea water samples, and will completely invalidate the results as the weight of salts left on the filter during drying can easily exceed that of the suspended particulate matter.

Although turbidity purports to measure approximately the same water quality property as TSS, the latter is more useful because it provides an actual weight of the particulate material present in the sample. In water quality monitoring situations, a series of more labor-intensive TSS measurements will be paired with relatively quick and easy turbidity measurements to develop a site-specific correlation. Once satisfactorily established, the correlation can be used to estimate TSS from more frequently made turbidity measurements, saving time and effort. Because turbidity readings are somewhat dependent on particle size, shape, and color, this approach requires calculating a correlation equation for each location. Further, situations or conditions that tend to suspend larger particles through water motion (e.g., increase in a stream current or wave action) can produce higher values of TSS not necessarily accompanied by a corresponding increase in turbidity. This is because particles above a certain size (essentially anything larger than silt) are not measured by a bench turbidity meter (they settle out before the reading is taken), but contribute substantially to the TSS value.

Issues

Although TSS appears to be a straightforward measure of particulate weight obtained by separating particles from a water sample using a filter, it suffers as a defined quantity from the fact that particles occur in nature in essentially a continuum of sizes. At the lower end, TSS relies on a cutoff established by properties of the filter being used. At the upper end, the cut-off should be the exclusion of all particulates too large to be "suspended" in water. However, this is not a fixed particle size but is dependent upon the energetics of the situation at the time of sampling: moving water suspends larger particles than does still water. Usually it is the case that the additional suspended material caused by the movement of the water is of interest.

These problems in no way invalidate the use of TSS; consistency in method and technique can overcome short-comings in most cases. But comparisons between studies may require a careful review of the methodologies used to establish that the studies are in fact measuring the same thing.

TSS in mg/L can be calculated as:

> (dry weight of residue and filter - dry weight of filter alone, in grams)/ mL of sample * 1,000,000.

Reasons for Caring about TSS

- Sediment enters our waterways through a wide range of methods; however, it is most often attributed to urbanization and agriculture. By removing trees and other natural vegetation, the soil structure is disturbed, which leads to increased erosion and sedimentation.

- Suspended solids enter Green Bay primarily through the Fox River and the smaller streams within the Fox River Basin.

- The suspended solids load at the mouth of the Fox River is approximately 151,915 tons/year (Total Maximum Daily Load and Watershed Management Plan for the lower Fox River Basin and Lower Green Bay, 2012 and State of the Bay, 2013), which is equivalent to 416 tons or roughly 27 standard dump trucks of sediment per day. However, almost 70 percent of the sediment is delivered within a short window (13-15 days) during the spring.

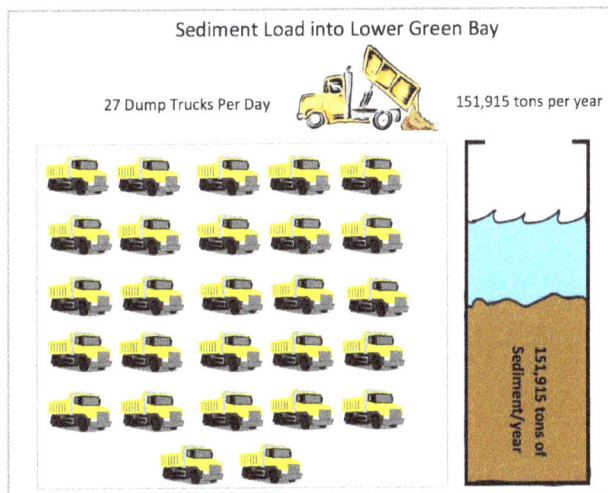

Sediment Load into Lower Green Bay

27 Dump Trucks Per Day   151,915 tons per year

151,915 tons of Sediment/year

Annual sediment load into Green Bay from the three major river basins (Lower Fox River, Upper Fox River, and Wolf River) 151,915 tons, which is equivalent to 416 tons per day or approximately 27 standard dump trucks (carrying 15.5 tons) per day (TMDL and Watershed Management Plan for the lower Fox River Basin and Lower Green Bay).

Impacts of Suspended Solids

- **Dead Zones:** Increased sediment reduces water clarity and therefore, less light is able to penetrate the waters' surface and reach plants underwater. Reduced sunlight, reduces photosynthesis activity and oxygen production, which can create hypoxic (low-oxygen) conditions and can contribute to dead zones.

- **Biological:** Increased sedimentation can have several negative impacts on the biology of an aquatic system.

  - Reduced water clarity decreases visibility for fish and diving birds.

  - Suspended solids can clog fish gills and cause increased levels of stress, which is often already amplified due to hypoxic conditions.

  - Suspended solids that settle at the bottom of a river or lake can bury fish eggs, fish nursery areas, and invertebrate habitat, which can negatively impact the food chain.

- **Economics:** Suspended solids settle at the bottom of a waterbody, creating a shallow waterway. Due to the shipping industry, which navigates through Green Bay and the Fox River, increased sedimentation has had significant financial ramifications. Maintaining shipping channels by dredging the sediment is a costly and continuous project.

References

- Franson, Mary Ann Standard Methods for the Examination of Water and Wastewater 14th edition (1975) APHA, AWWA & WPCF ISBN 0-87553-078-8

- Muller, Mathieu; Bouguelia, Sihem; Goy, Romy-Alice; Yoris, Alison; Berlin, Jeanne; Meche, Perrine; Rocher, Vincent; Mertens, Sharon; Dudal, Yves (2014). "International cross-validation of a BOD5 surrogate". Environmental Science and Pollution Research: 1–4. doi:10.1007/s11356-014-3202-3

- In-Situ Inc. Method 1002-8-2009 Dissolved Oxygen Measurement by Optical Probe, In-Situ Inc., 221 E Lincoln Ave., Ft. Collins, CO 80524 "Archived copy". Archived from the original on 2010-01-22. Retrieved 2010-01-12

- Wan Osman, Wan Hasnidah; Sheik Abdullah, Siti Rozaimah; Mohamad, Abu Baker; Kadhum, Abdul Amir H.; Abd Rahman, Ramki (2013) [2013]. "Simultaneous removal of AOX and COD from real recycled paper wastewater using GAC-SBBR". Journal of Environmental Management. 121: 80–86. doi:10.1016/j.jenvman.2013.02.005

- Connor, Richard (2016). The United Nations World Water Development Report 2016: Water and Jobs, chapter 2: The Global Perspective on Water. Paris: UNESCO. p. 26. ISBN 978-92-3-100155-0

- Hileman, Bette; Long, Janice R.; Kirschner, Elisabeth M. (1994-11-21). "Chlorine Industry Running Flat Out Despite Persistent Health Fears". Chemical & Engineering News Archive. 72 (47): 12–26. doi:10.1021/cen-v072n047.p012. ISSN 0009-2347

- Hogan, C. Michael; Patmore, Leda C.; Seidman, Harry (August 1973). "Statistical Prediction of Dynamic Thermal Equilibrium Temperatures using Standard Meteorological Data Bases". EPA. Retrieved 2016-02-15

Chapter 3

Wastewater Treatment Techniques

The treatment of wastewater is an important aspect of sanitation. It includes the management of human and solid waste, industrial wastewater and sewage treatment, stormwater management, etc. The aim of this chapter is to explore the different techniques of treatment of wastewater such as coagulation water treatment, secondary and tertiary wastewater treatment, use of API oil-water separator, septic tank, etc.

Wastewater Treatment

Wastewater treatment is the process of converting wastewater – water that is no longer needed or is no longer suitable for use – into bilge water that can be discharged back into the environment. It's formed by a number of activities including bathing, washing, using the toilet, and rainwater runoff. Wastewater is full of contaminants including bacteria, chemicals and other toxins. Its treatment aims at reducing the contaminants to acceptable levels to make the water safe for discharge back into the environment.

There are two wastewater treatment plants namely chemical or physical treatment plant, and biological wastewater treatment plant. Biological waste treatment plants use biological matter and bacteria to break down waste matter. Physical waste treatment plants use chemical reactions as well as physical processes to treat wastewater. Biological treatment systems are ideal for treating wastewater from households and business premises. Physical wastewater treatment plants are mostly used to treat wastewater from industries, factories and manufacturing firms. This is because most of the wastewater from these industries contains chemicals and other toxins that can largely harm the environment.

Step by Step Wastewater Treatment Process

The following is a step by step process of how wastewater is treated:

Wastewater Collection

This is the first step in waste water treatment process. Collection systems are put in place by municipal administration, home owners as well as business owners to ensure that all the wastewater is collected and directed to a central point. This water is then directed to a treatment plant using underground drainage systems or by exhauster tracks owned and operated by business people. The transportation of wastewater should however be done under hygienic conditions. The pipes or tracks should be leak proof and the people offering the exhausting services should wear protective clothing.

Odor Control

At the treatment plant, odor control is very important. Wastewater contains a lot of dirty substances that cause a foul smell over time. To ensure that the surrounding areas are free of the foul smell, odor treatment processes are initiated at the treatment plant. All odor sources are contained and treated using chemicals to neutralize the foul smell producing elements. It is the first wastewater treatment plant process and it's very important.

Screening

This is the next step in wastewater treatment process. Screening involves the removal of large objects for example nappies, cotton buds, plastics, diapers, rags, sanitary items, nappies, face wipes, broken bottles or bottle tops that in one way or another may damage the equipment. Failure to observe this step, results in constant machine and equipment problems. Specially designed equipment is used to get rid of grit that is usually washed down into the sewer lines by rainwater. The solid wastes removed from the wastewater are then transported and disposed off in landfills.

Primary Treatment

This process involves the separation of macrobiotic solid matter from the wastewater. Primary treatment is done by pouring the wastewater into big tanks for the solid matter to settle at the surface of the tanks. The sludge, the solid waste that settles at the surface of the tanks, is removed by large scrappers and is pushed to the center of the cylindrical tanks and later pumped out of the tanks for further treatment. The remaining water is then pumped for secondary treatment.

Secondary Treatment

Also known as the activated sludge process, the secondary treatment stage involves adding seed sludge to the wastewater to ensure that is broken down further. Air is first pumped into huge aeration tanks which mix the wastewater with the seed sludge which is basically small amount of sludge, which fuels the growth of bacteria that uses oxygen and the growth of other small microorganisms that consume the remaining organic matter. This process leads to the production of large particles that settle down at the bottom of the huge tanks. The wastewater passes through the large tanks for a period of 3-6 hours.

Bio-solids Handling

The solid matter that settle out after the primary and secondary treatment stages are directed to digesters. The digesters are heated at room temperature. The solid wastes are then treated for a month where they undergo anaerobic digestion. During this process, methane gases are produced and there is a formation of nutrient rich bio-solids which are recycled and dewatered into local firms. The methane gas formed is usually used as a source of energy at the treatment plants. It can be used to produce electricity in engines or to simply drive plant equipment. This gas can also be used in boilers to generate heat for digesters.

Tertiary Treatment

This stage is similar to the one used by drinking water treatment plants which clean raw water for drinking purposes. The tertiary treatment stage has the ability to remove up to 99 percent of the impurities from the wastewater. This produces effluent water that is close to drinking water quality. Unfortunately, this process tends to be a bit expensive as it requires special equipment, well trained and highly skilled equipment operators, chemicals and a steady energy supply. All these are not readily available.

Disinfection

After the primary treatment stage and the secondary treatment process, there are still some diseases causing organisms in the remaining treated wastewater. To eliminate them, the wastewater must be disinfected for at least 20-25 minutes in tanks that contain a mixture of chlorine and sodium hypochlorite. The disinfection process is an integral part of the treatment process because it guards the health of the animals and the local people who use the water for other purposes. The effluent (treated waste water) is later released into the environment through the local water ways.

Sludge Treatment

The sludge that is produced and collected during the primary and secondary treatment processes requires concentration and thickening to enable further processing. It is put into thickening tanks that allow it to settle down and later separates from the water. This process can take up to 24 hours. The remaining water is collected and sent back to the huge aeration tanks for further treatment. The sludge is then treated and sent back into the environment and can be used for agricultural use.

Industrial Wastewater Treatment

Industrial wastewater treatment is applicable to most manufacturing, mining, energy and petro-chemical companies aiming to decrease their liquid discharge into the environment. This benefits the environment and, importantly, helps companies save money by re-using large water volumes in their industrial processes.

The Potential of Industrial Wastewater Treatment

Environmental compliance

Veolia aims to help industrial clients comply with region-specific regulations pertaining to industrial wastewater treatment standards.

Process Water Re-use

By recycling water within a closed industrial wastewater treatment system, companies can achieve savings by feeding less clean water into a plant, and can reduce carbon and other pollutants entering the environment. Industrial wastewater treatment and re-use also help to free up potable water for distribution to communities.

Industrial Process Water from Seawater

With industry-leading desalination techniques, Veolia can turn high-salinity seawater into suitable for use in industrial processes. It is a well-established process of removing salts from water to produce process water using technologies like reverse osmosis, multiple effect distillation, multistage flash and hybrid systems that use thermal and reverse osmosis.

Industrial Wastewater Treatment for Human Consumption

Cutting-edge technologies are able to transform industrial wastewater into a fresh water resource for human consumption through various water purification technologies. Veolia uses leading water purification technologies and, worldwide, Veolia's global company has built more than 5 000 water purification plants.

Value Recovery from Industrial Wastewater

Industrial wastewater treatment technologies are able to extract valuable substances from process water before discharge or re-use within a plant, achieving greater sustainability and lower carbon footprints. By acting sustainably, customers recognise the value of water, help to save energy and produce it, and optimise costs. Reducing your carbon footprint helps create responsible value and a whole lot more.

Sources of Industrial Wastewater

Battery Manufacturing

Battery manufacturers specialize in fabricating small devices for electronics and portable equipment (e.g., power tools), or larger, high-powered units for cars, trucks and other motorized vehicles. Pollutants generated at manufacturing plants includes cadmium, chromium, cobalt, copper, cyanide, iron, lead, manganese, mercury, nickel, oil & grease, silver and zinc.

Organic Chemicals Manufacturing

The specific pollutants discharged by organic chemical manufacturers vary widely from plant to plant, depending on the types of products manufactured, such as bulk organic chemicals, resins, pesticides, plastics, or synthetic fibers. Some of the organic compounds that may be discharged are benzene, chloroform, napthalene, phenols, toluene and vinyl chloride. Biochemical oxygen demand (BOD), which is a gross measurement of a range of organic pollutants, may be used to gauge the effectveness of a biological wastewater treatment system, and is used as a regulatory parameter in some discharge permits. Metal pollutant discharges may include chromium, copper, lead, nickel and zinc.

Electric Power Plants

Fossil-fuel power stations, particularly coal-fired plants, are a major source of industrial wastewater. Many of these plants discharge wastewater with significant levels of metals such as lead, mercury, cadmium and chromium, as well as arsenic, selenium and nitrogen compounds (nitrates and nitrites). Wastewater streams include flue-gas desulfurization, fly ash, bottom ash and flue gas mercury control. Plants with air pollution controls such as wet scrubbers typically transfer the captured pollutants to the wastewater stream.

Ash ponds, a type of surface impoundment, are a widely used treatment technology at coal-fired plants. These ponds use gravity to settle out large particulates (measured as total suspended solids) from power plant wastewater. This technology does not treat dissolved pollutants. Power stations use additional technologies to control pollutants, depending on the particular wastestream in the plant. These include dry ash handling, closed-loop ash recycling, chemical precipitation, biological treatment (such as an activated sludge process), membrane systems, and evaporation-crystallization systems. Technological advancements in ion exchange membranes and electrodialysis systems has enabled high efficiency treatment of flue-gas desulfurization wastewater to meet recent EPA discharge limits. The treatment approach is similar for other highly scaling industrial wastewaters.

Food Industry

Wastewater generated from agricultural and food operations has distinctive characteristics that set it apart from common municipal wastewater managed by public or private sewage treatment plants throughout the world: it is biodegradable and non-toxic, but has high concentrations of biochemical oxygen demand (BOD) and suspended solids (SS). The constituents of food and agriculture wastewater are often complex to predict, due to the differences in BOD and pH in effluents from vegetable, fruit, and meat products and due to the seasonal nature of food processing and post-harvesting.

Processing of food from raw materials requires large volumes of high grade water. Vegetable washing generates waters with high loads of particulate matter and some dissolved organic matter. It may also contain surfactants.

Animal slaughter and processing produces very strong organic waste from body fluids, such as blood, and gut contents. This wastewater is frequently contaminated by significant levels of antibiotics and growth hormones from the animals and by a variety of pesticides used to control external parasites.

Processing food for sale produces wastes generated from cooking which are often rich in plant organic material and may also contain salt, flavourings, colouring material and acids or alkali. Very significant quantities of oil or fats may also be present.

Food processing activities such as plant cleaning, material conveying, bottling, and product washing create wastewater. Many food processing facilities require on-site treatment before operational wastewater can be land applied or discharged to a waterway or a sewer system. High suspended solids levels of organic particles increase Biochemical Oxygen Demand (BOD) can result in significant sewer surcharge fees. Sedimentation, wedgewire screening, or rotating belt filtration (microscreening) are commonly used methods to reduce suspended organic solids loading prior to discharge.

Iron and Steel Industry

The production of iron from its ores involves powerful reduction reactions in blast furnaces. Cooling waters are inevitably contaminated with products especially ammonia and cyanide. Production of coke from coal in coking plants also requires water cooling and the use of water in by-products separation. Contamination of waste streams includes gasification products such as benzene, naphthalene, anthracene, cyanide, ammonia, phenols, cresols together with a range of more complex organic compounds known collectively as polycyclic aromatic hydrocarbons (PAH).

The conversion of iron or steel into sheet, wire or rods requires hot and cold mechanical transformation stages frequently employing water as a lubricant and coolant. Contaminants include hydraulic oils, tallow and particulate solids. Final treatment of iron and steel products before onward sale into manufacturing includes *pickling* in strong mineral acid to remove rust and prepare the surface for tin or chromium plating or for other surface treatments such as galvanisation or painting. The two acids commonly used are hydrochloric acid and sulfuric acid. Wastewaters include acidic rinse waters together with waste acid. Although many plants operate acid recovery plants (particularly those using hydrochloric acid), where the mineral acid is boiled away from the iron salts, there remains a large volume of highly acid ferrous sulfate or ferrous chloride to be disposed of. Many steel industry wastewaters are contaminated by hydraulic oil, also known as *soluble oil*.

Mines and Quarries

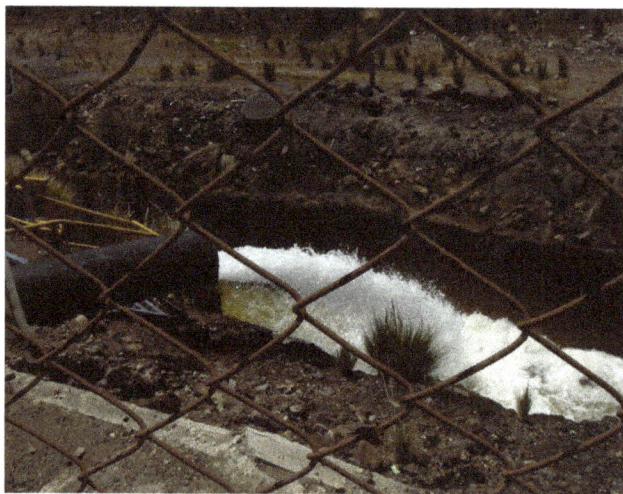

Mine wastewater effluent in Peru, with neutralized pH from tailing runoff.

The principal waste-waters associated with mines and quarries are slurries of rock particles in water. These arise from rainfall washing exposed surfaces and haul roads and also from rock washing and grading processes. Volumes of water can be very high, especially rainfall related arisings on large sites. Some specialized separation operations, such as coal washing to separate coal from native rock using density gradients, can produce wastewater contaminated by fine particulate haematite and surfactants. Oils and hydraulic oils are also common contaminants.

Wastewater from metal mines and ore recovery plants are inevitably contaminated by the minerals present in the native rock formations. Following crushing and extraction of the desirable materials, undesirable materials may enter the wastewater stream. For metal mines, this can include unwanted metals such as zinc and other materials such as arsenic. Extraction of high value metals such as gold and silver may generate slimes containing very fine particles in where physical removal of contaminants becomes particularly difficult.

Additionally, the geologic formations that harbour economically valuable metals such as copper and gold very often consist of sulphide-type ores. The processing entails grinding the rock into fine particles and then extracting the desired metal(s), with the leftover rock being known as tailings. These tailings contain a combination of not only undesirable leftover metals, but also sulphide components which eventually form sulphuric acid upon the exposure to air and water that inevitably occurs when the tailings are disposed of in large impoundments. The resulting acid mine drainage, which is often rich in heavy metals (because acids dissolve metals), is one of the many environmental impacts of mining.

Nuclear Industry

The waste production from the nuclear and radio-chemicals industry is dealt with as *Radioactive waste*.

Petroleum Refining and Petrochemicals

Pollutants discharged at petroleum refineries and petrochemical plants include conventional pollutants (biochemical oxygen demand, oil and grease, suspended solids), ammonia, chromium, phenols and sulfides.

Pulp and Paper Industry

Effluent from the pulp and paper industry is generally high in suspended solids and BOD. Plants that bleach wood pulp for paper making may generate chloroform, dioxins (including 2,3,7,8-TCDD), furans, phenols and chemical oxygen demand (COD). Stand-alone paper mills using imported pulp may only require simple primary treatment, such as sedimentation or dissolved air flotation. Increased BOD or COD loadings, as well as organic pollutants, may require biological treatment such as activated sludge or upflow anaerobic sludge blanket reactors. For mills with high inorganic loadings like salt, tertiary treatments may be required, either general membrane treatments like ultrafiltration or reverse osmosis or treatments to remove specific contaminants, such as nutrients.

Textile Dyeing

Textile dyeing plants generate wastewater that contain synthetic and natural dyestuff, gum thickener (guar) and various wetting agents, pH buffers and dye retardants or accelerators. Following treatment with polymer-based flocculants and settling agents, typical monitoring parameters include BOD, COD, color (ADMI), sulfide, oil and grease, phenol, TSS and heavy metals (chromium, zinc, lead, copper).

Industrial Oil Contamination

Industrial applications where oil enters the wastewater stream may include vehicle wash bays, workshops, fuel storage depots, transport hubs and power generation. Often the wastewater is discharged into local sewer or trade waste systems and must meet local environmental specifications. Typical contaminants can include solvents, detergents, grit. lubricants and hydrocarbons.

Water Treatment

Many industries have a need to treat water to obtain very high quality water for demanding purposes such pure chemical synthesis or boiler feed water. Many water treatment produce organic and mineral sludges from filtration and sedimentation. Ion exchange using natural or synthetic resins removes calcium, magnesium and carbonate ions from water, typically replacing them with sodium, chloride, hydroxyl and/or other ions. Regeneration of ion exchange columns with strong acids and alkalis produces a wastewater rich in hardness ions which are readily precipitated out, especially when in admixture with other wastewater constituents.

Wood Preserving

Wood preserving plants generate conventional and toxic pollutants, including arsenic, COD, copper, chromium, abnormally high or low pH, phenols, oil & grease, and suspended solids.

Wool Processing

Insecticide residues in fleeces are a particular problem in treating waters generated in wool processing. Animal fats may be present in the wastewater, which if not contaminated, can be recovered for the production of tallow or further rendering.

Treatment of Industrial Wastewater

The various types of contamination of wastewater require a variety of strategies to remove the contamination.

Brine Treatment

Brine treatment involves removing dissolved salt ions from the waste stream. Although similarities to seawater or brackish water desalination exist, industrial brine treatment may contain unique combinations of dissolved ions, such as hardness ions or other metals, necessitating specific processes and equipment.

Brine treatment systems are typically optimized to either reduce the volume of the final discharge for more economic disposal (as disposal costs are often based on volume) or maximize the recovery of fresh water or salts. Brine treatment systems may also be optimized to reduce electricity consumption, chemical usage, or physical footprint.

Brine treatment is commonly encountered when treating cooling tower blowdown, produced water from steam assisted gravity drainage (SAGD), produced water from natural gas extraction such as coal seam gas, frac flowback water, acid mine or acid rock drainage, reverse osmosis reject, chlor-alkali wastewater, pulp and paper mill effluent, and waste streams from food and beverage processing.

Brine treatment technologies may include: membrane filtration processes, such as reverse osmosis; ion exchange processes such as electrodialysis or weak acid cation exchange; or evaporation processes, such as brine concentrators and crystallizers employing mechanical vapour recompression and steam.

Reverse osmosis may not be viable for brine treatment, due to the potential for fouling caused by hardness salts or organic contaminants, or damage to the reverse osmosis membranes from hydrocarbons.

Evaporation processes are the most widespread for brine treatment as they enable the highest degree of concentration, as high as solid salt. They also produce the highest purity effluent, even distillate-quality. Evaporation processes are also more tolerant of organics, hydrocarbons, or hardness salts. However, energy consumption is high and corrosion may be an issue as the prime mover is concentrated salt water. As a result, evaporation systems typically employ titanium or duplex stainless steel materials.

Brine Management

Brine management examines the broader context of brine treatment and may include consideration of government policy and regulations, corporate sustainability, environmental impact, recycling, handling and transport, containment, centralized compared to on-site treatment, avoidance and reduction, technologies, and economics. Brine management shares some issues with leachate management and more general waste management.

Solids Removal

Most solids can be removed using simple sedimentation techniques with the solids recovered as slurry or sludge. Very fine solids and solids with densities close to the density of water pose special problems. In such case filtration or ultrafiltration may be required. Although, flocculation may be used, using alum salts or the addition of polyelectrolytes. Wastewater from industrial food processing often requires on-site treatment before it can be discharged to prevent or reduce sewer surcharge fees. The type of industry and specific operational practices determine what types of wastewater is generated and what type of treatment is required. Reducing solids such as waste product, organic materials, and sand is often a goal of industrial wastewater treatment. Some common ways to reduce solids include primary sedimentation (clarification), Dissolved Air Flotation or (DAF), belt filtration (microscreening), and drum screening.

Oils and Grease Removal

The effective removal of oils and grease is dependent on the characteristics of the oil in terms of its suspension state and droplet size, which will in turn affect the choice of separator technology. Oil in industrial waste water may be free light oil ,heavy oil, which tends to sink, and emulsified oil, often referred to as soluble oil. Emulsified or soluble ooils will typically required "cracking" to free the oil from its emulsion. In most cases this is achieved by lowering the pH of the water matrix.

Most separator technologies will have an optimum range of oil droplet sizes that can be effectively treated.

Analysing the oily water to determine droplet size can be performed with a video particle analyser. Each separator technology will have its own performance curve outlining optimum performance based on oil droplet size. the most common separators are gravity tanks or pits, API oil-water separators or plate packs, chemical treatment via DAFs, centrifuges, media filters and hydrocyclones.

API Separators

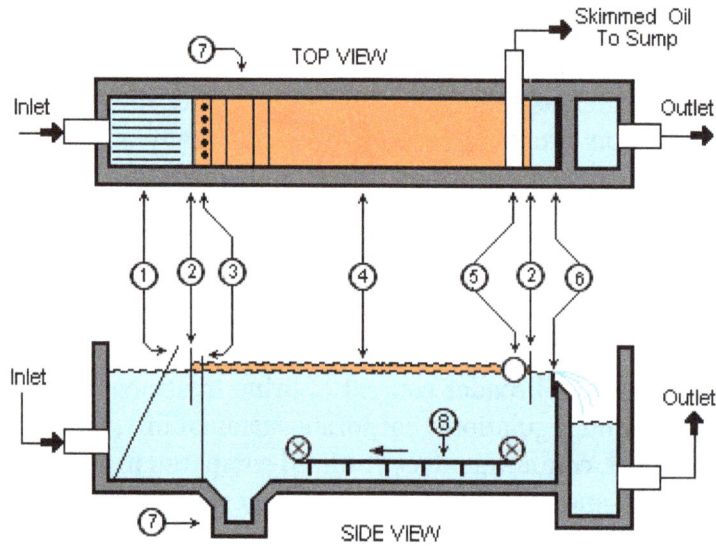

1 Trash trap (inclined rods)
2 Oil retention baffles
3 Flow distributors (vertical rods)
4 Oil layer
5 Slotted pipe skimmer
6 Adjustable overflow weir
7 Sludge sump
8 Chain and flight scraper

A typical API oil-water separator used in many industries

Many oils can be recovered from open water surfaces by skimming devices. Considered a dependable and cheap way to remove oil, grease and other hydrocarbons from water, oil skimmers can sometimes achieve the desired level of water purity. At other times, skimming is also a cost-efficient method to remove most of the oil before using membrane filters and chemical processes. Skimmers will prevent filters from blinding prematurely and keep chemical costs down because there is less oil to process.

Because grease skimming involves higher viscosity hydrocarbons, skimmers must be equipped with heaters powerful enough to keep grease fluid for discharge. If floating grease forms into solid clumps or mats, a spray bar, aerator or mechanical apparatus can be used to facilitate removal.

However, hydraulic oils and the majority of oils that have degraded to any extent will also have a soluble or emulsified component that will require further treatment to eliminate. Dissolving or emulsifying oil using surfactants or solvents usually exacerbates the problem rather than solving it, producing wastewater that is more difficult to treat.

The wastewaters from large-scale industries such as oil refineries, petrochemical plants, chemical plants, and natural gas processing plants commonly contain gross amounts of oil and suspended solids. Those industries use a device known as an API oil-water separator which is designed to separate the oil and suspended solids from their wastewater effluents. The name is derived from the fact that such separators are designed according to standards published by the American Petroleum Institute (API).

The API separator is a gravity separation device designed by using Stokes Law to define the rise velocity of oil droplets based on their density and size. The design is based on the specific gravity difference between the oil and the wastewater because that difference is much smaller than the specific gravity difference between the suspended solids and water. The suspended solids settles to the bottom of the separator as a sediment layer, the oil rises to top of the separator and the cleansed wastewater is the middle layer between the oil layer and the solids.

Typically, the oil layer is skimmed off and subsequently re-processed or disposed of, and the bottom sediment layer is removed by a chain and flight scraper (or similar device) and a sludge pump. The water layer is sent to further treatment for additional removal of any residual oil and then to some type of biological treatment unit for removal of undesirable dissolved chemical compounds.

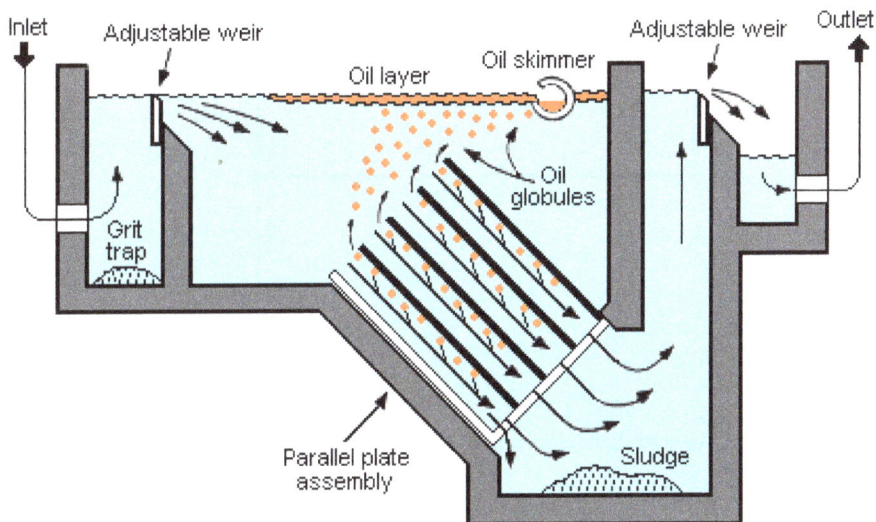

A typical parallel plate separator

Parallel plate separators are similar to API separators but they include tilted parallel plate assemblies (also known as parallel packs). The parallel plates provide more surface for suspended oil droplets to coalesce into larger globules. Such separators still depend upon the specific gravity between the suspended oil and the water. However, the parallel plates enhance the degree of oil-water separation. The result is that a parallel plate separator requires significantly less space than a conventional API separator to achieve the same degree of separation.

Hydrocyclone Oil Separators

Hydrocyclone oil separators operate on the process where wastewater enters the cyclone chamber and is spun under extreme centrifugal forces more than 1000 times the force of gravity. This force causes the water and oil droplets to separate. The separated oil is discharged from one end of the cyclone where treated water is discharged through the opposite end for further treatment, filtration or discharge.

Removal of Biodegradable Organics

Biodegradable organic material of plant or animal origin is usually possible to treat using extended conventional sewage treatment processes such as activated sludge or trickling filter. Problems can arise if the wastewater is excessively diluted with washing water or is highly concentrated such as undiluted blood or milk. The presence of cleaning agents, disinfectants, pesticides, or antibiotics can have detrimental impacts on treatment processes.

Activated Sludge Process

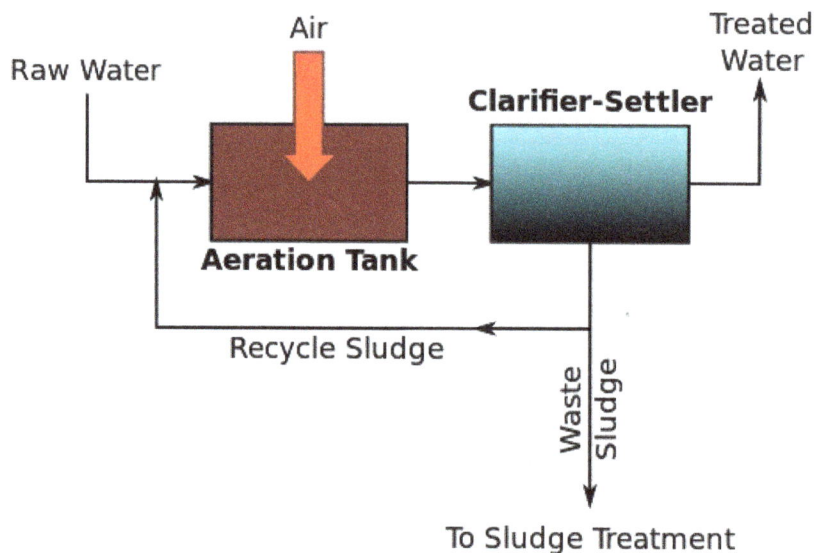

A generalized diagram of an activated sludge process.

Activated sludge is a biochemical process for treating sewage and industrial wastewater that uses air (or oxygen) and microorganisms to biologically oxidize organic pollutants, producing a waste sludge (or floc) containing the oxidized material. In general, an activated sludge process includes:

- An aeration tank where air (or oxygen) is injected and thoroughly mixed into the wastewater.

- A settling tank (usually referred to as a clarifier or "settler") to allow the waste sludge to settle. Part of the waste sludge is recycled to the aeration tank and the remaining waste sludge is removed for further treatment and ultimate disposal.

Trickling Filter Process

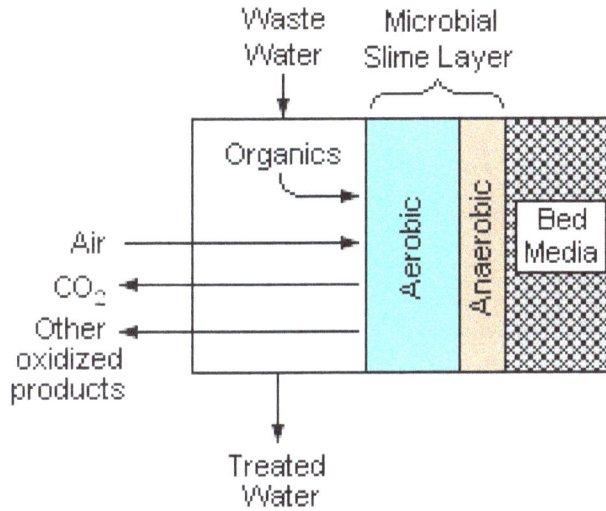

A schematic cross-section of the contact face of the bed media in a trickling filter

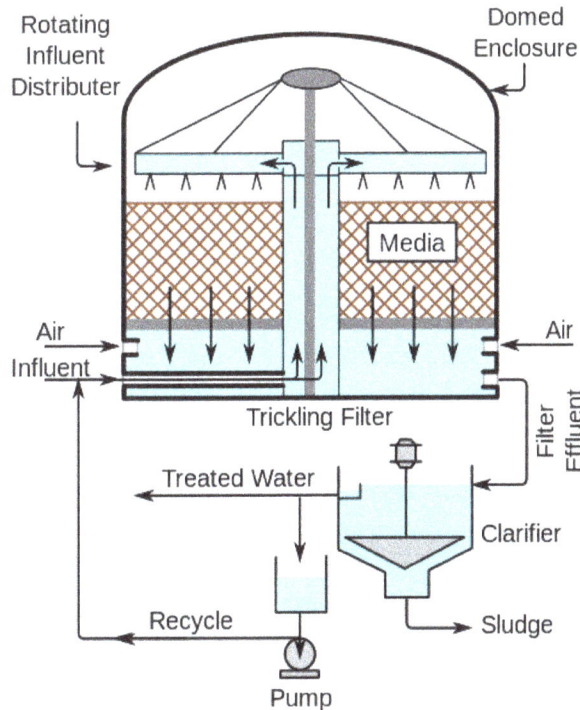

A typical complete trickling filter system

A trickling filter consists of a bed of rocks, gravel, slag, peat moss, or plastic media over which wastewater flows downward and contacts a layer (or film) of microbial slime covering the bed media. Aerobic conditions are maintained by forced air flowing through the bed or by natural convection of air. The process involves adsorption of organic compounds in the wastewater by the microbial slime layer, diffusion of air into the slime layer to provide the oxygen required for the biochemical oxidation of the organic compounds. The end products include carbon dioxide gas, water and other products of the oxidation. As the slime layer thickens, it becomes difficult for the air to penetrate the layer and an inner anaerobic layer is formed.

The fundamental components of a complete trickling filter system are:

- A bed of filter medium upon which a layer of microbial slime is promoted and developed.

- An enclosure or a container which houses the bed of filter medium.

- A system for distributing the flow of wastewater over the filter medium.

- A system for removing and disposing of any sludge from the treated effluent.

The treatment of sewage or other wastewater with trickling filters is among the oldest and most well characterized treatment technologies.

A trickling filter is also often called a *trickle filter*, *trickling biofilter*, *biofilter*, *biological filter* or *biological trickling filter*.

Removal of Other Organics

Synthetic organic materials including solvents, paints, pharmaceuticals, pesticides, products from coke production and so forth can be very difficult to treat. Treatment methods are often specific to the material being treated. Methods include advanced oxidation processing, distillation, adsorption, vitrification, incineration, chemical immobilisation or landfill disposal. Some materials such as some detergents may be capable of biological degradation and in such cases, a modified form of wastewater treatment can be used.

Removal of Acids and Alkalis

Acids and alkalis can usually be neutralised under controlled conditions. Neutralisation frequently produces a precipitate that will require treatment as a solid residue that may also be toxic. In some cases, gases may be evolved requiring treatment for the gas stream. Some other forms of treatment are usually required following neutralisation.

Waste streams rich in hardness ions as from de-ionisation processes can readily lose the hardness ions in a buildup of precipitated calcium and magnesium salts. This precipitation process can cause severe *furring* of pipes and can, in extreme cases, cause the blockage of disposal pipes. A 1-metre diameter industrial marine discharge pipe serving a major chemicals complex was blocked by such salts in the 1970s. Treatment is by concentration of de-ionisation waste waters and disposal to landfill or by careful pH management of the released wastewater.

Removal of Toxic Materials

Toxic materials including many organic materials, metals (such as zinc, silver, cadmium, thallium, etc.) acids, alkalis, non-metallic elements (such as arsenic or selenium) are generally resistant to biological processes unless very dilute. Metals can often be precipitated out by changing the pH or by treatment with other chemicals. Many, however, are resistant to treatment or mitigation and may require concentration followed by landfilling or recycling. Dissolved organics can be *incinerated* within the wastewater by the advanced oxidation process.

Smart Capsules

Molecular encapsulation is a technology that has the potential to provide a system for the recyclable removal of lead and other ions from polluted sources. Nano-, micro- and milli- capsules, with sizes in the ranges 10 nm-1μm,1μm-1mm and >1mm, respectively, are particles that have an active reagent (core) surrounded by a carrier (shell).There are three types of capsule under investigation: alginate-based capsules, carbon nanotubes, polymer swelling capsules. These capsules provide a possible means for the remediation of contaminated water.

Sedimentation Wastewater Treatment

Sedimentation is the most common physical unit operation in wastewater treatment, more so in primary treatment where sedimentation is the workhorse of the treatment. The term sedimentation is also called settling in some literature. Sedimentation is, in a nutshell, a process by which the suspended solids, which have higher densities than that of water, are re-moved from wastewater by the action of gravity in the bottom of the settling tank or basin (also called a clarifier) within a reasonable period of time. Sedimentation basins are usually rectangular or circular with a radial or upward water flow pattern. Sedimentation is not limited to primary treatment; there is also secondary sedimentation by which settleable solids in the biological secondary treatment processes are removed. For example, recovery of activated sludge for recycling is achieved with secondary sedimentation.

A photo of a dissolved air flotation system.

In a typical wastewater tretment plant, the wastewater stream exiting from screening devices (and after flotation basins) then enters the second section of the primary treatment of wastewater treatment or sedimentation tanks/basins. Here, the sludge (the organic portion of the sewage) settles out of the wastewater and is pumped out of the tanks. Some of the water is removed in a step called thickening and then the sludge is processed in large tanks called digesters.

Sedimentation uses gravitational force to separate unstable and destabilized suspended solids from wastewater. It is based on the density differ ence between the bulk of the liquid and the solids. Stabilized solids such as colloids can be destabilized with flocculants. Sedimentation is a very important primary treatment process; it is, however, also used in the biological treatment, such as activated sludge and trickling filters for solid removal. The settling characteristics of the solids are determined by the types of the settling solids and their concentrations. Sedimentation has four distinct types of settling:

- Discrete settling (Type I), which is settling of a dilute suspension of solids that do not aggregate.

- Flocculent settling (Type II), which is settling of the particulates that aggregate among themselves and/or with added flocculants to form larger particulates and therefore results in faster settling. The sedimentation operation in a typical primary treatment of wastewater operates in this mode.

- Zone settling (Type III, also called hindered settling), which occurs when particulates adhere together, forming a mass that settles as a blanket with a distinguishable interface with the liquid above it.

- Compression zone (Type IV), which occurs when sinking particulates accumulate at the bottom of the sedimentation tank/basin, forming a compressed structure that supports the weight of the particulates that settled in the bottom of the tank/basin.

Although sedimentation basins in primary treatment are characterized by Type II flocculant settling, each of these zones have different characteristics that warrant further analysis.

Discrete Settling (Type I)

The settling of nonaggregated solids in a dilute suspension can be described by its settling velocity of individual particulates. In a settling tank/ basin, the settling of a discrete particle is not affected by the other particles and is only a function of the fluid property and the characteristics of the particle; this may be further depicted in Figure when the movement of the particle of interest is subject to the combined effect of the gravitational force downward and the bulk flow toward the outlet:

vt = (tank depth)/(residence time)

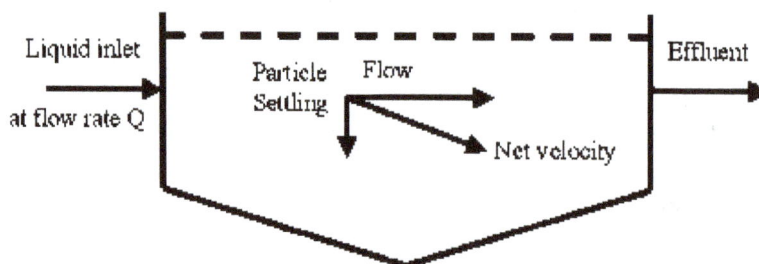

A schematic diagram of discrete settling.

or mathematically: vt = H/t

H is the depth of the sedimentation tank and t is the residence time of the particle in the liquid in the tank. If we assume the residence time of the liquid is the residence time of the liquid in the tank.

A is the cross-sectional area of the tank and Q is the overall volumetric flow rate through the tank. Here, Q/A is the overflow rate of the liquid passing through the tank.

The terminal velocity of the particle is equal to the overflow rate of the tank.

For readers who have some exposure to fluid mechanics, the above derivation might strike them as being suspicious, since the terminal velocity of a discrete particle in the liquid.

where d is the size (equivalent diameter of the particle), g is the gravitational acceleration, ^ is the viscosity of the liquid, and (ps — pl) is the density difference between the particle and the liquid.

This suggests that the terminal velocity is a function of size and the density of the particle, which is nowhere to be inferred. Note that we have assumed that the residence time of the particle in the tank is the same as that of the bulk liquid or the equivalent depth for the particle settling in the tank is the same as the depth of the tank. The overflow rate of the tank as the terminal velocity of a particle in the tank represents the critical velocity of an ideal particle in an ideal settling tank assuming:

- The number of ideal discrete particles and the velocity vectors of the liquid are uniformly distributed.

- The liquid flows in the tank as an ideal slug.

- Any particle reaching the bottom of the tank is effectively removed (no resuspension) (Canale and Borchardt, 1972).

Any ideal particles having termial velocity v (average velocity among all particles present) greater than vt is 100% removed from the settling tank/basin. For those particles with less than vt average terminal velocity, the portion of the particles removed in the tank is equal to v/vt.

In reality, the discrete settling is more likely associated with settling of hard particulates with high density and size such as grit and sand. This is a rare type of particulate in a typical food wastewater stream, but it may occur in some sources of agricultural wastewater that is subject to intrusion of soil and dirt.

Flocculent Settling (Type II)

Flocculent settling is used in primary clarifiers and the upper zones of secondary clarifiers. In the case of flocculent settling, the particles in the relatively dilute suspension coalesce or flocculate to form larger particles or aggregates during settling, thus increasing the mass of settling solids as well as the settling velocity (and removal rate). In many food wastewater treatment situations, except very dilute ones, suspended solids cannot be described as discrete particles of known specific gravity (a quantity that is the ratio of density of particle to density of water). In general, larger particles settle faster and have a greater tendency to collide with other slower-settling particles, resulting in formation of larger particles in a quiescent body of water. However, the wind, hydrodynamic shear, and hydraulic disturbance all contribute to further contacts among particles in the tank.

Sample Ports 60cm Apart

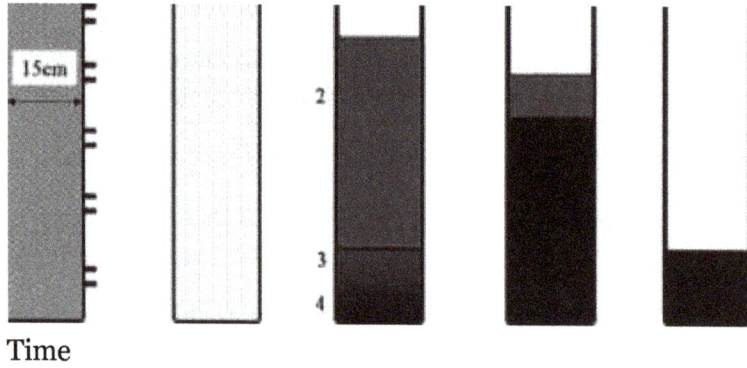

Time

A diagram of settling column and zoning settling process.

Furthermore, the greater the depth of the tank, the higher the frequency of collisons among particles will be during settling. Therefore, the flocculent settling is dependent on the properties of particles and the liquid as well as depth of the settling tank/basin. The settled solids in the bottom of the tank are usually promptly removed, so the greater rate of settling as a result of aggregation of individual particles translates into a greater rate of solid removal from the wastewater. Evaluation of a wastewater stream slated for a sedimention tank or basin is carried out using a settling column, as depicted in Figure. The laboratory of the settling column is about 15 cm (6 in) in diameter and 305 cm (10 ft) tall, and it has several sampling ports 61 cm (2 ft) apart. The settling evaluation is conducted by first placing a known quantity of wastewater sample in the column. The uniformity of particle size from top to bottom of the column in the beginning of the evaluation and the temperature of the liquid throughout the evaluation should be accomplished. The wastewater containing suspended solids is allowed to settle under quiescent conditions; small samples of suspension at different ports with preset depths are drawn and concentrations of particles in the samples are determined over preset time intervals. The frac tion removal of the particles is calculated for each sample analyzed and is plotted against time and depth. The fraction of the particles removed at each depth is constructed as curve lines called isoconcentration lines, as those illustrated in Figure. These lines represent the most efficient particle removal loci for a given removal rate. The ratio of the depth to time is the average settling (terminal) velocity of the particles under a given percent removal.

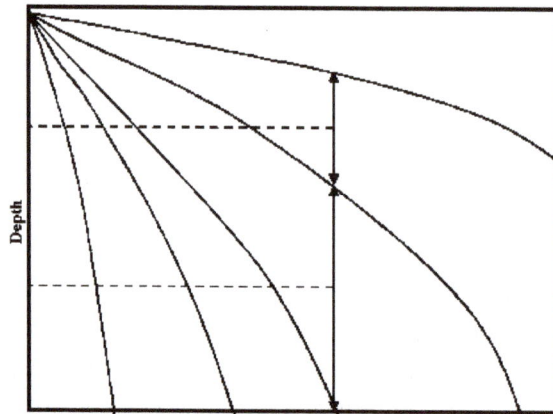

Time

A fraction of removal of flocculating particles at each depth.

Zone Settling (Type III)

Zone settling, also called hindered settling, acquires its name from the fact that aggregated particulates of a concentrated suspension (activated sludge or flocculated colloids) in the sedimentation basin tend to form a massive blanketlike suspension with a distinct interface. Zone settling is mainly used in secondary clarifiers. Many wastewater treatment process designers use a batch settling test to determine the interface.

Physicochemical Wastewater Treatment Processes 65 Compression Zone (Type IV)

Compression settling involves the highest concentration of suspended solids and occurs in the lower reaches of clarifiers. The particles settle by compressing the mass of the particles below. Compression occurs not only in the lower zones of secondary clarifiers but also in sludge thickening tanks.

Basics

Suspended solids (or SS), is the mass of dry solids retained by a filter of a given porosity related to the volume of the water sample. This includes particles 10 μm and greater.

Colloids are particles of a size between 0.001 μm and 1 μm depending on the method of quantification. Because of Brownian motion and electrostatic forces balancing the gravity, they are not likely to settle naturally.

The limit sedimentation velocity of a particle is its theoretical descending speed in clear and still water. In settling process theory, a particle will settle only if :-

1. In a vertical ascending flow, the ascending water velocity is lower than the limit sedimentation velocity.

2. In a longitudinal flow, the ratio of the length of the tank to the height of the tank is higher than the ratio of the water velocity to the limit sedimentation velocity.

Removal of suspended particles by sedimentation depends upon the size and specific gravity of those particles. Suspended solids retained on a filter may remain in suspension if their specific gravity is similar to water while very dense particles passing through the filter may settle. Settleable solids are measured as the visible volume accumulated at the bottom of an Imhoff cone after water has settled for one hour.

Gravitational theory is employed, alongside the derivation from Newton's second law and the Navier–Stokes equations.

Stokes' law explains the relationship between the settling rate and the particle diameter. Under specific conditions, the particle settling rate is directly proportional to the square of particle diameter and inversely proportional to liquid viscosity.

The settling velocity, defined as the residence time taken for the particles to settle in the tank, enables the calculation of tank volume. Precise design and operation of a sedimentation tank is of high importance in order to keep the amount of sediment entering the diversion system to a minimum threshold by maintaining the transport system and stream stability to remove the sediment diverted from the system. This is achieved by reducing stream velocity as low as possible for the longest period of time possible. This is feasible by widening the approach channel and lowering its floor to reduce flow velocity thus allowing sediment to settle out of suspension due to gravity. The settling behavior of heavier particulates is also affected by the turbulence.

Designs

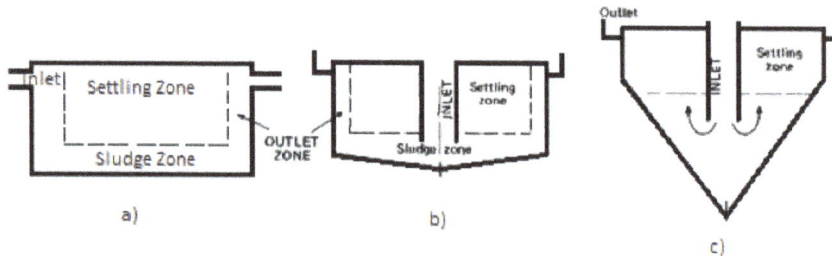

Different clarifier designs

Although sedimentation might occur in tanks of other shapes, removal of accumulated solids is easiest with conveyor belts in rectangular tanks or with scrapers rotating around the central axis of circular tanks. Settling basins and clarifiers should be designed based on the settling velocity of the smallest particle to be theoretically 100% removed. The overflow rate is defined as:

Overflow rate (V_o) = Flow of water (Q (cubic metre per second)) /(Surface area of settling basin (A))(m^2)

In many countries this value is named as surface loading in m3/h per m2. Overflow rate is often used for flow over an edge (for example a weir) in the unit m3/h per m.

The unit of overflow rate is usually meters (or feet) per second, a velocity. Any particle with settling velocity (*Vs*) greater than the overflow rate will settle out, while other particles will settle in the ratio Vs/Vo. There are recommendations on the overflow rates for each design that ideally take into account the change in particle size as the solids move through the operation:

- Quiescent zones: 9.4 mm (0.031 ft) per second

- Full-flow basins: 4.0 mm (0.013 ft) per second

- Off-line basins: 0.46 mm (0.0015 ft) per second

However, factors such as flow surges, wind shear, scour, and turbulence reduce the effectiveness of settling. To compensate for these less than ideal conditions, it is recommended doubling the area calculated by the previous equation. It is also important to equalize flow distribution at each point across the cross-section of the basin. Poor inlet and outlet designs can produce extremely poor flow characteristics for sedimentation.

Settling basins and clarifiers can be designed as long rectangles (Figure. a), that are hydraulically more stable and easier to control for large volumes. Circular clarifiers (Figure. b) work as a common thickener (without the usage of rakes), or as upflow tanks (Figure. c).

Sedimentation efficiency does not depend on the tank depth. If the forward velocity is low enough so that the settled material does not re-suspend from the tank floor, the area is still the main parameter when designing a settling basin or clarifier, taking care that the depth is not too low.

Assessment of Main Process Characteristics

Settling basins and clarifiers are designed to retain water so that suspended solids can settle. By sedimentation principles, the suitable treatment technologies should be chosen depending on the specific gravity, size and shear resistance of particles. Depending on the size and density of particles, and physical properties of the solids, there are four types of sedimentation processes:

- Type 1 – Dilutes, non-flocculent, free-settling (every particle settles independently.)

- Type 2 – Dilute, flocculent (particles can flocculate as they settle).

- Type 3 – Concentrated suspensions, zone settling, hindered settling (sludge thickening).

- Type 4 – Concentrated suspensions, compression (sludge thickening).

Different factors control the sedimentation rate in each.

Settling of Discrete Particles

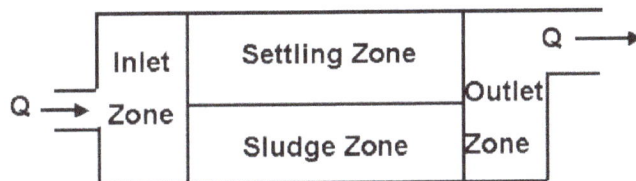

The four functional zones of a continuous flow settling basin

Unhindered settling is a process that removes the discrete particles in a very low concentration without interference from nearby particles. In general, if the concentration of the solutions is lower than 500 mg/L total suspended solids, sedimentation will be considered discrete. Concentrations of raceway effluent total suspended solids (TSS) in the west are usually less than 5 mg/L net. TSS concentrations of off-line settling basin effluent are less than 100 mg/L net. The particles keep their size and shape during discrete settling, with an independent velocity. With such low concentrations of suspended particles, the probability of particle collisions is very low and consequently the rate of floculation is small enough to be neglected for most calculations. Thus the surface area of the settling basin becomes the main factor of sedimentation rate. All continuous flow settling basins are divided into four parts: inlet zone, settling zone, sludge zone and outlet zone.

In the inlet zone, flow is established in a same forward direction. Sedimentation occurs in the settling zone as the water flow towards to outlet zone. The clarified liquid is then flow out from outlet zone. Sludge zone: settled will be collected here and usually we assume that it is removed from water flow once the particles arrives the sludge zone.

In an ideal rectangular sedimentation tank, in the settling zone, the critical particle enters at the top of the settling zone, and the settle velocity would be the smallest value to reach the sludge zone, and at the end of outlet zone, the velocity component of this critical particle are Vs, the settling velocity in vertical direction and Vh in horizontal direction.

From Figure, the time needed for the particle to settle;

$$t_o = H/V_p = L/V_s \quad (3)$$

Since the surface area of the tank is WL, and Vs = Q/WL, Vh = Q/WH, where Q is the flow rate and W, L, H is the width, length, depth of the tank.

According to Equation, this also is a basic factor that can control the sedimentation tank performance which called overflow rate.

Equation also tell us that the depth of sedimentation tank is independent to the sedimentation efficiency, only if the forward velocity is low enough to make sure the settled mass would not suspended again from the tank floor.

Settlement of Flocculent Particles

In a horizontal sedimentation tank, some particles may not follow the diagonal line in Figure, while settling faster as they grow. So this says that particles can grow and develop a higher settling velocity if a greater depth with longer retention time. However, the collision chance would be even greater if the same retention time were spread over a longer, shallower tank. In fact, in order to avoid hydraulic short-circuiting, tanks usually are made 3–6 m deep with retention times of a few hours.

Zone-settling Behaviour

As the concentration of particles in a suspension is increased, a point is reached where particles are so close together that they no longer settle independently of one another and the velocity fields of the fluid displaced by adjacent particles, overlap. There is also a net upward flow of liquid displaced by the settling particles. This results in a reduced particle-settling velocity and the effect is known as hindered settling.

There is a common case for hindered settling occurs. the whole suspension tends to settle as a 'blanket' due to its extremely high particle concentration. This is known as zone settling, because it is easy to make a distinction between several different zones which separated by concentration discontinuities. Figure represents a typical batch-settling column tests on a suspension exhibiting zone-settling characteristics. There is a clear interface near the top of the column would be formed to separating the settling sludge mass from the clarified supernatant as long as leaving such a suspension to stand in a settling column. As the suspension settles, this interface will move down at the same speed. At the same time, there is an interface near the bottom between that settled suspension and the suspended blanket. After settling of suspension is complete, the bottom interface would move upwards and meet the top interface which moves downwards.

Compression Settling

The settling particles can contact each other and arise when approaching the floor of the sedimentation tanks at very high particle concentration. So that further settling will only occur in adjust matrix as the sedimentation rate decreasing. This is can be illustrated by the lower region of the zone-settling diagram. In Compression zone, the settled solids are compressed by gravity (the weight of solids), as the settled solids are compressed under the weight of overlying solids, and water is squeezed out while the space gets smaller.

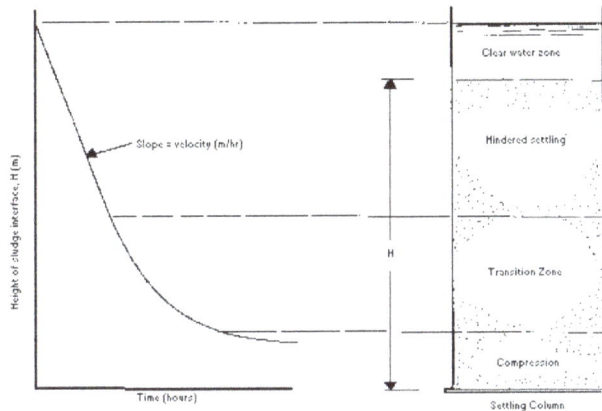

Typical batch-settling column test on a suspension
exhibiting zone-settling characteristics

Sedimentation Tank

Sedimentation tank, also called settling tank or clarifier, component of a modern system of water supply or wastewater treatment. A sedimentation tank allows suspended particles to settle out of water or wastewater as it flows slowly through the tank, thereby providing some degree of purification. A layer of accumulated solids, called sludge, forms at the bottom of the tank and is periodically removed. In drinking-water treatment, coagulants are added to the water prior to sedimentation in order to facilitate the settling process, which is followed by filtration and other treatment steps. In modern sewage treatment, primary sedimentation must be followed by secondary treatment (e.g., trickling filter or activated sludge) to increase purification efficiencies. Sedimentation is usually preceded by treatment using bar screens and grit chambers to remove large objects and coarse solids.

Different Types of Sedimentation

There are various different types of sedimentation:

- Granular particles will each settle separately and will each be subject to a constant settling speed;

- Particles that are more or less flocculated will be of different sizes and, therefore, subject to variable sedimentation rates. At low concentration levels, the settling speed increases as the floc increases in size through collision with other particles; this is termed flocculent settling.

At higher concentration levels, the abundance of floc and its interactions will lead to an overall sedimentation which is most frequently characterised by a clearly defined interface between the sludge mass and the supernatant liquid: this is hindered settling where the rate will be optimum within a certain area of concentration, above which, we talk of hindered settling.

Granular Particle Sedimentation

This is the simplest case and the only one that can be easily described by equations.

Theory: Fluid at Rest

When a granular particle remains in a liquid at rest, it will be affected by a driving force F_M (gravity minus the Archimedes thrust) and a resisting force F_T (fluid drag) created by viscosity and inertia forces:

$$F_M = g \cdot v \cdot \Delta\rho$$

$$F_T = \frac{C \cdot s \cdot \rho_F \cdot V^2}{2}$$

ρ_p, ρ_L: density of the granular particle and of the fluid, and $\Delta\rho = \rho_p - \rho_L$,

d,s,v : diameter, projected surface area (master couple: $\dfrac{\pi d^2}{4}$ for a sphere)

and volume of the granular particle.

V: particle settling rate,
g: gravity acceleration,
C: drag coefficient (adimensional),
Very quickly, a ($F_M = F_T$) balance will become established and the sedimentation of the particle treated like a sphere will occur at a constant rate V, such that:

$$V_0^2 = \frac{4}{3} g \frac{d}{C} \frac{\Delta\rho}{\rho_F}$$

Hydraulic Conditions

The value of C, the drag coefficient, is defined by the disturbance which is in turn based on the settling speed. This disturbance is characterised by the Reynolds grain number established by:

$$Re = \frac{\rho_F \cdot V \cdot d}{\mu}$$

Re = adimensional.
where μ = dynamic viscosity.

When Re is low, the viscosity forces are far greater than the inertia forces. When Re is high, the viscosity forces are negligible.

The drag coefficient is provided by:

$$C = a \cdot Re^{-n}$$

Table provides the various values for a, n and C according to the Reynolds number.

Re	Condition	a	n	C	Formula
$10^{-4} < Re < 1$	Laminar	24	1	$24 \cdot Re^{-1}$	Stokes
$1 < Re < 10^3$	Intermediate	18.5	0.6	$18.5 \cdot Re^{-0.6}$	Allen
$10^3 < Re < 2 \cdot 10^5$	Turbulent	0.44	0	0.44	Newton

Table 9. Hydraulic conditions

These formulas constitute the basis used for calculating grain movement in a fluid and are used for sedimentation (granular solids in a liquid, water drops in air), in upward motion (air bubbles in water, oil drops in water), for centrifugation, and for fluidisation.

Under laminar conditions, the Stokes law applied to a spherical particle will give:

$$V_0 = \frac{g}{18 \cdot \mu} \Delta \rho \cdot d^2$$

The aggregation phenomena that cause growth will, therefore, very rapidly increase the sedimentation rate.

Under transitory conditions, the Allen law also gives a rising rate based on particle size; however, this rate rises far more slowly because:

$$V_0^{1.4} = \frac{g}{13.875 \cdot \mu^{0.6}} \frac{\Delta \rho}{\rho_F^{0.4}} d^{16}$$

$$\text{then } V_0 = k \cdot d^{1,143}$$

Sphericity Factor

This factor Ψ is provided by:

$$\Psi = \frac{\text{volume of the sphere having the same surface}}{\text{grain volume}}$$

In the preceding operations, we then need to replace C by C' = ΨC and the Stokes law will be written:

$$V_0 = \frac{g}{18 \cdot \mu} \Delta\rho \cdot d^2$$

Table illustrates the considerable effect this factor has on "flat" materials.

Values for ψ	
Sand	2
Carbon	2.25
Talc	
Gypsum	4
Graphite lamella	22
Mica	170

The considerable effect this factor has on "flat" materials

Recovery Conditions

Let us consider a rectangular sedimentation tank that has a length L, a vertical section S = H·ℓ (where H is the water depth and ℓthe width) and a horizontal section S_H = L· ℓ, evenly crossed by a throughput Q that either ascends vertically or travels horizontally; the following conditions will apply if the tank is to screen out a granular particle settling at a rate V_0 in still water:

Vertical Upflow Sedimentation

Particles having a sedimentation rate that is greater than the liquid's upwards velocity will be screened out. This phenomenon is written:

$$V_0 > V_{asc} = \frac{Q}{S_H}$$

Q = liquid flow rate.
S_H = settling tank free surface area.

horizontal flow sedimentation

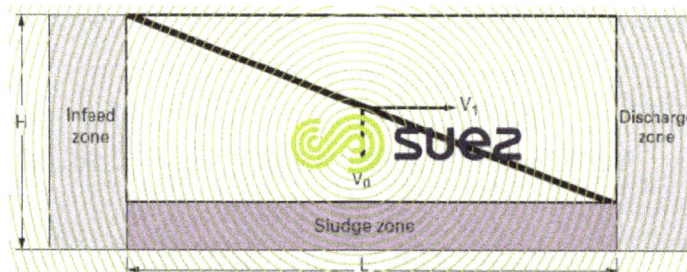

Schematic diagram of a horizontal flow sedimentation
application (granular particles)

The velocity of a particle entering into the tank through the top will have two components:

V_1: fluid horizontal velocity equal to Q/S.

V_0: vertical settling speed provided by Stokes law.

This particle will be retained in the tank when the time it takes for the particle to touch the tank floor (or to penetrate into the sludge zone)

$$t_1 = \frac{H}{V_0}$$ is less than the water's contact time in the sedimentation tank $$t_2 = \frac{L}{V_1} = \frac{L \cdot S}{Q}.$$

Then:

$$\frac{H}{V_0} < \frac{L \cdot S}{Q} \quad \text{or} \quad V_0 > \frac{H \cdot Q}{L \cdot S} = \frac{H \cdot Q}{L \cdot \ell \cdot H} = \frac{Q}{S_H} = V_H$$

V_H: Hazen velocity (or hydraulic loading on the surface) similar to Vasc in the preceding example and expressed in $m^3 \cdot (h \cdot m^2)^{-1}$ or $m \cdot h^{-1}$.

It should be noted that V_H is independent of tank depth.

All particles having sedimentation rates above V_H will, theoretically, be eliminated. However, if the water infeed is distributed over the entire depth of the tank, some of the particles that have a sedimentation rate V below the Hazen velocity will also be retained within the V/V_H ratio whereas these particles would not have been retained in a upward flow sedimentation tank.

In theory, for identical horizontal surface areas, a horizontal flow sedimentation tank can thus be used to separate a greater number of particles.

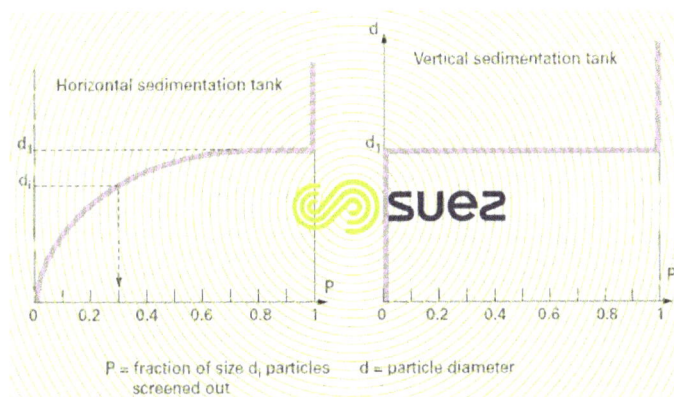

P = fraction of size d_i particles   d = particle diameter
screened out

Effectiveness comparison between horizontal and
upward flow sedimentation (granular particles)

In practice, this difference will be attenuated and even reversed for the following reasons associated with horizontal flow sedimentation :

- Hydraulic distribution problems in the vertical plane both at the inlet and at the outlet of a structure;

- Sludge accumulation and collection, reducing the available section;

- In a circular, horizontal flow sedimentation tank, the horizontal component of the particle velocity (V_1) decreases from the centre outwards and the particle will adopt a curvilinear trajectory.

Flocculent Settling of Flocculated Particles

During sedimentation, flocculation will continue to take place and particle sedimentation rate V_o will rise.

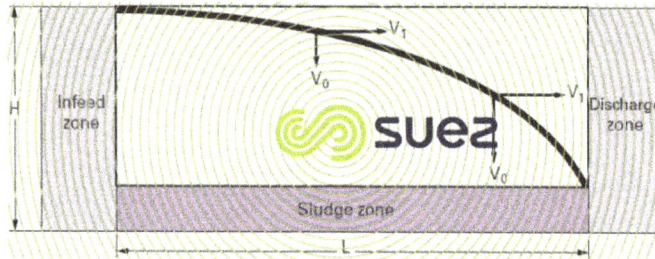

Schematic diagram of a horizontal flow sedimentation
application (flocculated particles)

This process takes place as soon as flocculated matter concentration rises above approximately 50 mg · L^{-1}.

Flocculent sedimentation effectiveness does not depend only on the hydraulic loading at the surface but also on contact time. There are no mathematical formulae available for calculating the sedimentation rate.

Only laboratory tests and graph methods can be used to ascertain this rate. Figure provides the results of one such test.

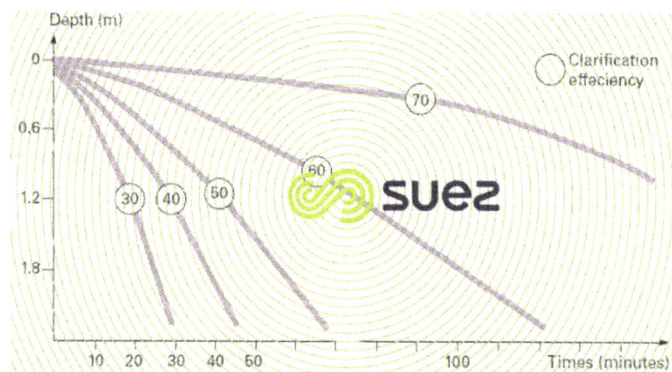

Eliminating flocculated particles by the flocculent sedimentation
method: relationship between time, effective depth and sedimentation performance

Hindered Settling of Flocculated Particles

When the flocculated particle concentration increases, the interaction between particles can no longer be ignored; they undergo "hindered" settling. Initially, this may cause flocculation and sedimentation to improve and then hindered beyond a certain critical concentration; we then talk of "hindered settling".

This is a phenomenon that is characteristic of activated sludge and flocculated suspensions when their concentration rises above approximately 500 mg · L⁻¹.

Visual Observation

When hindered settling is carried out in a tube of adequate height and diameter (at least a 1-litre test tube), we normally see the appearance of four zones.

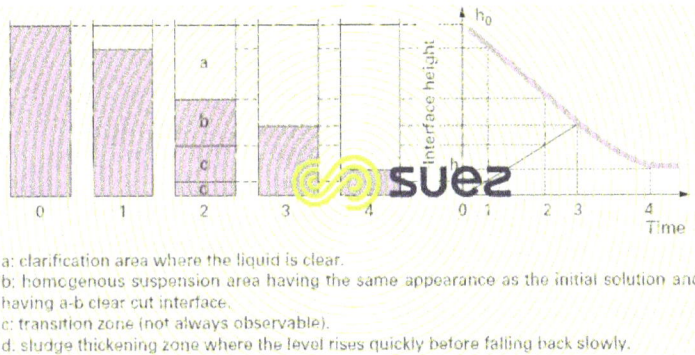

a: clarification area where the liquid is clear.
b: homogenous suspension area having the same appearance as the initial solution and having a-b clear cut interface.
c: transition zone (not always observable).
d. sludge thickening zone where the level rises quickly before falling back slowly.

Hindered settling: Kynch curve

After a certain state, zones b and c disappear; this is the critical point. The change in the a-b interface height as a function of time constitutes the Kynch curve.

Kynch Curve

Kynch's fundamental hypothesis is that the rate at which a particle falls depends exclusively on local particle concentration C Courbe de Kynch.

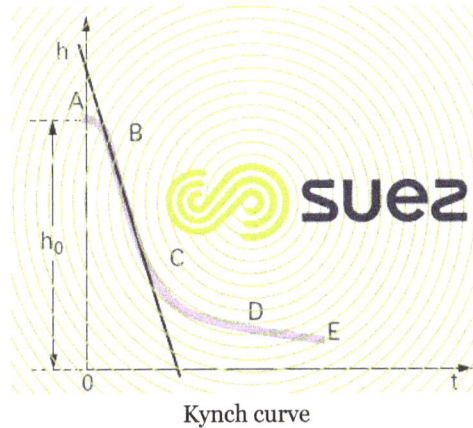

Kynch curve

The separation surface between A and B is more or less clear-cut: this is the flake coalescence phase. This phase does not always exist.

From B to C, a rectilinear part identifies a constant settling speed V (straight line slope). In the case of a tube of specific dimensions, V depends on the initial suspended solids concentration and on the suspension's flocculation properties. When the initial concentration C increases, the mass's sedimentation rate V decreases: e.g. in activated urban sludge where the suspended solids concentration goes from 1 to 4 g · L⁻¹, Vdecreases from 6 to 1.8 m · h⁻¹.

The CD section which is concave at the top, refers to a gradual decrease in the settling speed of the deposit's top layer.

From D, the flakes come into contact with each other and exert a compression effect on the lower layers.

Kynch's theory applies to section BC and CD that cover the main flocculated sludge sedimentation area.

Interpretation

Let us consider a suspension where its clarification does not include a coalescence phase, calculations show that:

- In the BOC triangle, concentration and settling speed remain constant and equal to the initial values found in B;

- In the COD triangle, the equal concentration curves are straight lines that cross through the origin; this means that, as soon as sedimentation commences, the layers that are the closest to the base of the tank will have to transit through every concentration from initial concentration to that applicable at point D, the start of compression.

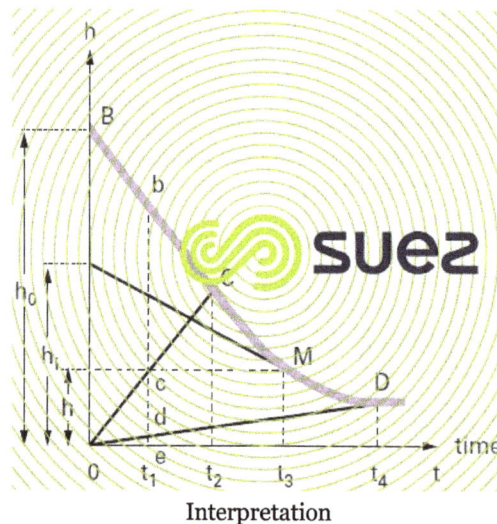

Interpretation

Consequently, the sludge medium that has a height eb at point in time t_1 will have three separate zones:

- An upper zone bc where settling speed and concentration remain uniform and will have retained their initial values V_0 and C;

- An intermediate zone cd where concentration increases gradually from c to d and the settling speed drops accordingly;

- A lower zone where sludge flakes are subject to compression.

In the medium considered at a point in time t_2, the upper zone disappears and, at point in time t_4, only the lower zone remains.

With regard to point M in section CD, there are two concentration levels:

C_M^i concentration at the interface,

C_M mean concentration.

According to the Kynch hypothesis:

$$C_M^I = C_0 \frac{h_o}{h_i}$$

Furthermore:

$$C_M^I = C_0 \frac{h_o}{h}$$

The three sections BC, CD and DE on the Kynch curve (figure 16) are used to dimension hindered settling tanks. The BC phase refers to the solids contact clarification area. The CD phase applies to structures where sludge concentration is targeted (units used for thickened sludge recirculation). The DE phase is used for sludge thickening.

Mohlman Index (Sludge Volume Index: Svi)

One specific point needs to be considered on the Kynch curve, that of the 30 minute abscissa: the Mohlman index I_M is widely used to define that clarification capacity of biological sludge and, therefore, for dimensioning its clarifier and even for initiating remedial action should bulking occur, see section biomass used in wastewater purification and chapter biological processes).

$$I_M = \frac{V}{M} \ (cm^3 \cdot g^{-1})$$

V: volume of sludge after 30 minute settling time (cm^3),
M: suspended solids present in this volume (g).

The drawback of the Mohlman index is that it is heavily dependent on sludge initial concentration. Therefore, methods have been put forward for establishing an independent index for this concentration and which is, therefore, exclusively characteristic of the state of the sludge in the plant concerned. This is:

Sludge Index (IB) or Diluted Mohlman Index (DSVI).

The Kynch curve is plotted from dilutions used to achieve a volume of sludge equal to approximately 200-250 mL per L after 30 minutes of settling time. The sludge index I_B used in France applies when this volume is between 100 and 300 mL; the DSVI used by the Anglo-Saxons applies when the volume is equal to between 150 and 250 mL. In both cases, it can be said that activated sludge has very good settling properties when its sludge index or its DSVI is between 50 and 100 $cm^3 \cdot g^{-1}$, normal settling properties between 100 and 200 and poor settling properties above 250.

Designing Settling Tanks

Two criteria are used for the purpose of calculating the surface area of a settling tank:

- The hydraulic surface loading that is characteristic of the volume of water that can be processed by the unit of surface area and unit of time ($m^3 \cdot m^{-2} \cdot h^{-1}$);

- The solids loading that is characteristic of suspended solids that will settle per unit of surface area and per unit of time ($kg \cdot m^{-2} \cdot h^{-1}$).

The most limiting of the two calculations will constitute the structures dimensioning parameter.

Influence of the Hydraulic Surface Loading

This loading is directly linked to "free or flocculated" suspended solids sedimentation rate and the preceding paragraph"different types of sedimentation" shows how this rate can be determined and how to calculate the ensuing minimum surface area.

Influence of Solids Loading

In flocculated particle hindered settling which includes thickening phenomena, mass flow rate is usually a determining factor when calculating settling tank surface area.

Let us consider a settling tank having a section S, fed with an incoming flow Q_E with a suspended solids concentration C_E; sludge is drawn off from the bottom of the settling tank at a flow rate of Q_S with a concentration C_S.

In the absence of chemical or biological reactions affecting suspended solids concentration, and if we consider a 100% elimination efficiency, we will have:

- Treated flow $Q = Q_E - Q_S$
- Materials balance $\cdot C_S = Q_E \cdot C_E$

or as solids loading:

$$\frac{Q_s \cdot C_s}{S} = \frac{Q_E \cdot C_E}{S}$$

The solids loading that will settle is provided by the Kynch curve. For a given point on the Kynch curve, having a concentration C_i, the sedimentation rate V_i is provided by the tangent at this point. The corresponding mass flow is $F_i = C_i \cdot V_i$.

To this mass flow F_i, we need to add the draw off mass flow F_s provided by $C_i V_s = Q_s/S$.

The total solids loading will be $F = C_i \cdot V_i + C_i \cdot V_s$.

Figure illustrates the changes undergone by these different flows. Flow F has a minimum F_L which is associated with a critical concentration C_L, which requires the settling tank to have a minimum section of S_m such that:

$$S_m = \frac{Q_E \cdot C_E}{F_L}$$

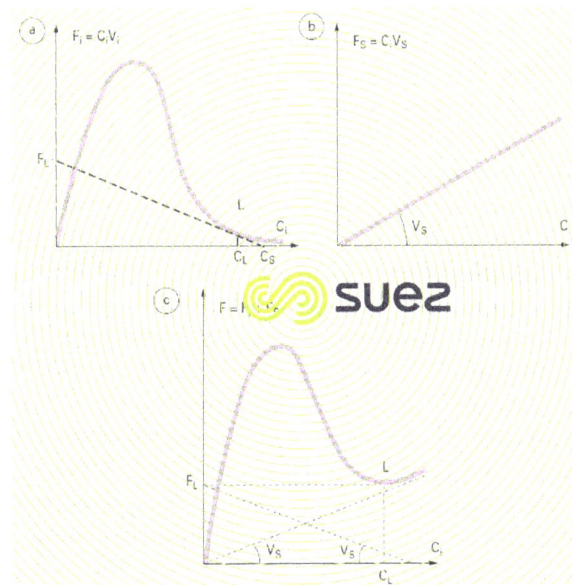

Solids loading curves

This specific point L can be determined direct on the solids loading curve F (figure 18 c) using:

$$\left(\frac{dF}{dC_i} \right)_L = \left(\frac{dF_i}{dC_i} \right)_L + V_s = 0$$

Point L is, therefore, the point on flow curve Fi where the slope of the tangent is equal in absolute value to the draw-off speed V_s. These results can be expressed in a variety of ways by considering the Kynch curve. Limit flow F_L at point L is provided by:

$$F_L = C_L(V_L + V_s) = C_L \left(V_L + \frac{Q_E \cdot C_E}{S} \times \frac{1}{C_S} \right)$$

V_L being the sedimentation rate at point L.

Therefore, for settling to be feasible:

$$\frac{Q_E \cdot C_E}{S} < \frac{V_L}{\dfrac{1}{C_L} - \dfrac{1}{C_S}}$$

Settling Tank Structure

$$R_{e*} = \frac{V \cdot d_h}{\upsilon}$$

Re^* : Reynolds number,
V : water circulation velocity ($m \cdot s^{-1}$),
d_h : equivalent hydraulic diameter (m),
υ : water's kinematic viscosity ($m^2 \cdot s^{-1}$),
with:

$$d_h = 4 \frac{\text{wet surface}}{\text{wet perimeter}}$$

In practice, the ideal settling tank does not exist: turbulence will occur within the liquid, especially in the intake zone and wind can create waves on the surface of the liquid; convection currents created by local temperatures (sunshine) and density differentials will have an effect on clarification performance. Wherever possible, the aim must be to obtain a stable, laminar circulation characterised by an appropriate Reynolds number values established by:

Note: In the case of a full circular pipe, the hydraulic diameter will be the pipe's diameter.

In practice, conditions are deemed to be laminar when $Re^* < 800$.

Furthermore, the Froude number can be used to evaluate circulation process stability when the main influence on the flow is generated by gravity and inertia forces.

$$Fr = \frac{V^2}{g \cdot d_h}$$

The more stable the circulation, the more even the velocity distribution over the entire tank section and the better the separation. However, stable circulation can be characterised by high Froude numbers.

In practice, we can establish optimum H/L or H/R ratios where H is the wetted depth of rectangular settling tanks having a length L or of circular settling tanks having a radius R. Taking a 2-hour contact time, Schmidt-Bregas makes the following recommendations:

- Horizontal flow, rectangular settling tanks

$$\frac{1}{35} < \frac{H}{L} < \frac{1}{20}$$

- Circular settling tanks:

$$\frac{1}{8} < \frac{H}{R} < \frac{1}{6}$$

The shape of the structure, the design of the mechanism for supplying raw water and of the mechanism for collecting treated water, as well as the sludge removal mode will all have a significant impact on settling tank performance.

In the case of water or liquor that is heavily loaded with suspended solids, the "density currents" can produce velocity distributions that tend to cause suspended solids that have accumulated on the tank floor to rise in the direction of the recovery channel. This is, for instance, the case of conventional rectangular or circular settling tanks that are too long and that are used to clarify activated sludge.

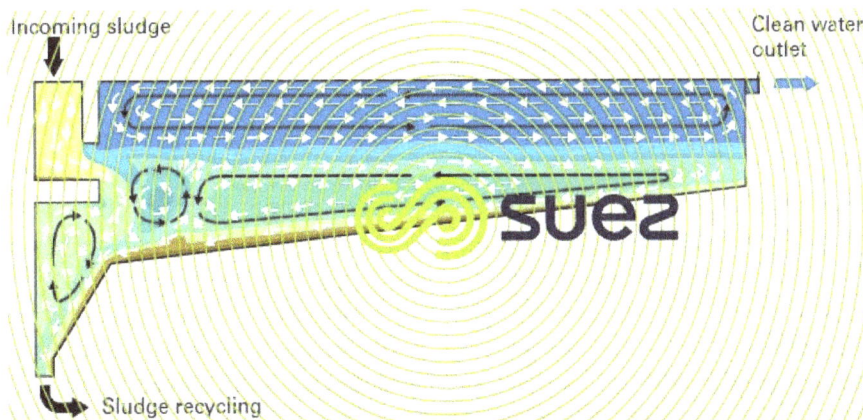

Density currents in a settling tank (CFD study)

Similarly, the effects of temperature (sunshine, water subject to rapid temperature fluctuations) and disturbances linked to changes in salinity (estuary water, IWW) will inject variable density water into the settling tank, creating convection currents, even completely overturning the water mass in the settling tank.

Clarifier

Clarifiers are required where ever the Suspended solids in raw/waste water are higher in concentration. Almost all treatment plant (Clarifiers) sedimentation tanks of circular or sometimes rectangular design.

Clarifiers work on the principle of gravity settling. The heavier suspended solids settle in the clarifier due to the quiescent conditions provided in the Clarification zone. The settled solids are swept to the centre well provided for collection of sludge with help of moving scraper blades.

Many a times the natural settling is enhanced by addition of coagulant & polyelectrolyte. The coagulant neutralizes the charges & agglomerates the suspended solids to form micro-floc. The polyelectrolyte brings together these micro-flocs binding them with long chains to make heavy floc which easily settles down. Most of the waste waters contain some scum material which does not settle down & needs to be collected on the surface of the clarifier. Hence a scum removal arrangement is normally provided. A simple clarifier is depicted in sketch below.

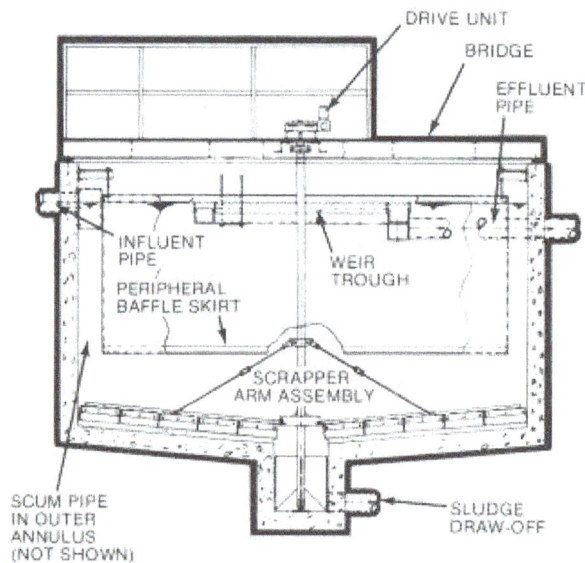

To facilitate the dosing of Chemicals as well as to ensure proper flocculation, many clarifiers are equipped with either separate flocculator or some clarifiers have a flocculating zone also.

Configurations

The clarifiers are designed based on the intended function & the space availability. In conventional waste water treatment plants there are normally two types:

   a. Primary Clarifier

   b. Secondary Clarifier

Specific applications require the following:

    a. Lamella Clarifier

    b. Solids Contact Clarifier

There are many design variation among the above type but available space & flow dictates whether it will be a circular or rectangular clarifier.

Circular Clarifier

The Clarifier design can be applied to water or wastewater treatment systems. It includes a larger influent well to provide the required flocculation time. Mixing is also added to achieve economical flocculation.

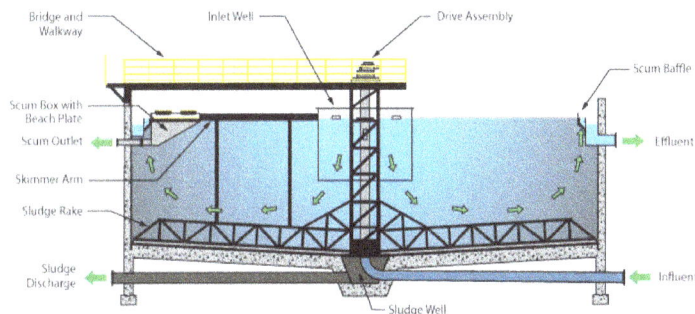

Designed to exacting specifications, the Clarifier is shop fabricated for bolted assembly, comprises of a reinforced influent well and features cast iron drive housings, deep scraper blades, adjustable squeegees and full surface adjustable skimmers.

Mechanical flocculation is proved by either concentric 'stacked' drives or independent mixers. Tanks start at ten feet in diameter and larger. Circular sludge collectors are available in either full or half bridge designs.

Rectangular Clarifier

The Chain & Scraper Sludge Collection System provides maximum sludge concentrations and scum/floating solid removal regardless of the size or application.

Rectangular clarification, a separation process commonly used in very large or space constrained municipal and industrial spaces, removes both settled and suspended solids from liquids.

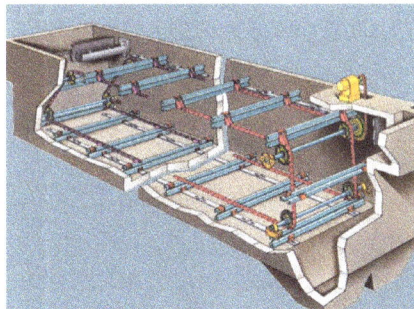

Suitable for primary, secondary, storm water collection and water plant service, Chain & Scraper is a high-quality option for both new and existing rectangular clarifier installations. To operate, flights mounted on two parallel strands of metallic or non-metallic chain scrape the settled solids along the tank floor to sludge hoppers. On the return run, the flights can skim the surface and concentrate the floating material at a scum removal device. Both three and two shaft designs are available if skimming is not a necessary function.

To assist the gravity settling many innovations are made in clarifier design.

A very popular design has Inclined plates (Lamella Clarifier) in the clarifying zone for rapid settling of flocs. The plates are placed at an angle (of 55-65°) & the water travels upwards. The solids (flocs) due to inherent weight cannot travel with same velocity & tend to settle on the plates provided. The slope of the plates ensures that the solids slide down to the collection chamber.

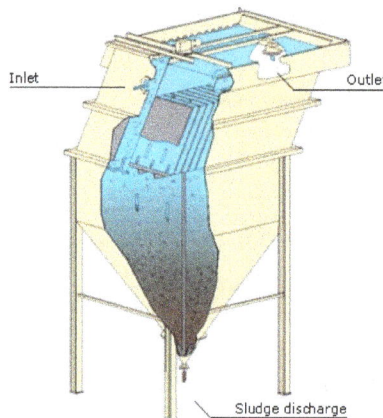

A schematic of Lamella Clarifier

Another design involves creating slow turbulence in the settled sludge itself. The upward moving sludge helps to bring down the settleable flocs faster to sludge zone.

Applications

Pretreatment

Before the water enters the clarifier, coagulation and flocculation reagents, such as polyelectrolytes and ferric sulfate, can be added. These reagents cause finely suspended particles to clump together and form larger and denser particles, called flocs, that settle more quickly and stably. This allows the separation of the solids in the clarifier to occur more efficiently and easily; aiding in the conservation of energy. Isolating the particle components first using these processes may reduce the volume of downstream water treatment processes like filtration.

Potable water treatment

Water being purified for human consumption, is treated with floculation reagents, then sent to the clarifier where removal of the flocculated coagulate occurs producing clarified water. The clarifier works by permitting the heavier and larger particles to settle to the bottom of the clarifier. The particles then form a bottom layer of sludge requiring regular removal and disposal. Clarified water then proceeds through several more steps before being sent for storage and use.

Waste water treatment

Sedimentation tanks have been used to treat wastewater for millennia.

Primary treatment of sewage is removal of floating and settleable solids through sedimentation. *Primary clarifiers* reduce the content of suspended solids and pollutants embedded in those suspended solids. Because of the large amount of reagent necessary to treat domestic wastewater, preliminary chemical coagulation and flocculation are generally not used, remaining suspended solids being reduced by following stages of the system. However, coagulation and flocculation can be used for building a compact treatment plant (also called a "package treatment plant"), or for further polishing of the treated water.

Sedimentation tanks called *secondary clarifiers* remove flocs of biological growth created in some methods of secondary treatment including activated sludge, trickling filters and rotating biological contactors.

Mining

Methods used to treat suspended solids in mining wastewater include sedimentation and floc blanket clarification and filtration. Sedimentation is used by Rio Tinto Minerals to refine raw ore into refined borates. After dissolving the ore, the saturated borate solution is pumped into a large settling tank. Borates float on top of the liquor while rock and clay settles to the bottom.

Technology

Rectangular sedimentation tanks with effluent weir structure visible above the fluid surface.

Drained circular sedimentation tank showing central inlet baffles on the right with solids scraper and skimmer arms visible under the rotating bridge.

Although sedimentation might occur in tanks of other shapes, removal of accumulated solids is easiest with conveyor belts in rectangular tanks or with scrapers rotating around the central axis of circular tanks. Mechanical solids removal devices move as slowly as practical to minimize resuspension of settled solids. Tanks are sized to give water an optimal residence time within the tank. Economy favors using small tanks; but if flow rate through the tank is too high, most particles will not have sufficient time to settle, and will be carried with the treated water. Considerable attention is focused on reducing water inlet and outlet velocities to minimize turbulence and promote effective settling throughout available tank volume. Baffles are used to prevent fluid velocities at the tank entrance from extending into the tank; and overflow weirs are used to uniformly distribute flow from liquid leaving the tank over a wide area of the surface to minimize resuspension of settling particles.

Tube Settlers

Tube settlers are commonly used in rectangular clarifiers to increase the settling capacity by reducing the vertical distance a suspended particle must travel. High efficiency tube settlers use a stack of parallel tubes, rectangles or flat pieces separated by a few inches (several centimeters) and sloping upwards in the direction of flow. This structure creates a large number of narrow parallel flow pathways encouraging uniform laminar flow as modeled by Stokes' law. These structures work in two ways:

1. They provide a very large surface area onto which particles may fall and become stabilized.

2. Because flow is temporarily accelerated between the plates and then immediately slows down, this helps to aggregate very fine particles that can settle as the flow exits the plates.

Structures inclined between 45° and 60° may allow gravity drainage of accumulated solids, but shallower angles of inclination typically require periodic draining and cleaning. Tube settlers may allow the use of a smaller clarifier and may enable finer particles to be separated with residence times less than 10 minutes. Typically such structures are used for difficult-to-treat waters, especially those containing colloidal materials.

Tube settlers capture the fine particles allowing the larger particles to travel to the bottom of the clarifier in a more uniform form. The fine particles then build up into a larger mass which then slides down the tube channels. The reduction in solids present in the outflow allows a reduction in the clarifier footprint when designing. Tubes made of PVC plastic are a minor cost in clarifier design improvements and may lead to an increase of operating rate of 2 to 4 times.

Operation

In order to maintain and promote the proper processing of a clarifier, it is important to remove any corrosive, reactive and polymerisable component first, or any material that may foul the outlet stream of water to avoid any unwanted side reactions, changes in the product or cause damage to any of the water treatment equipment. This is done by routine inspections and the frequent cleaning of the quiescent zones and the inlet and outlet areas of the clarifier in order to ascertain the extent of sediment build up and to clean and remove any scouring, litter, weeds or debris that may have accumulated over time.

Water being introduced into the clarifier should be controlled to reduce the velocity of the inlet flow. Reducing the velocity maximizes the hydraulic retention time inside the clarifier for sedimentation and helps to avoid excessive turbulence and mixing; thereby promoting the effective settling of the suspended particles. To further discourage the overt mixing within the clarifier and increase the retention time allowed for the particles to settle, the inlet flow should also be distributed evenly across the entire cross section of the settling zone inside the clarifier, where the volume is maintained at 37.7 percent capacity.

The sludge formed from the settled particles at the bottom of each clarifier, if left for an extended period of time, may become gluey and viscous, causing difficulties in its removal. This formation of sludge, promotes anaerobic conditions and a healthy environment for the growth of bacteria. This can cause the resuspension of particles by gases and the release of dissolved nutrients throughout the water fluid, reducing the effectiveness of the clarifier. Major health issues and problems can also occur further down the track of the water purification system or hinder the health of the fish found downstream of the clarifier.

Settling Basin

One of the methods implemented to address sediment loading, a settling basin or settling pond, have become widespread across the valley as a quick and simple way to remove sediments out of return flows. Settling basins simply slow down return flows, allowing sediments to settle out of the water before the water returns to the irrigation delivery system or other water body.

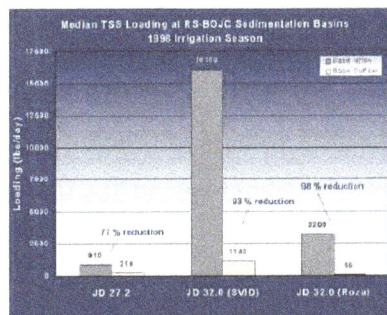

This graph shows the decrease in sediment loading over time in settling basins. Each basin's performance will vary depending on detention time, volume, and size. These variables will provide for varying results, but the bottom line, as evidence in the graph above, settling basins work very well removing sedimentation from return flows.

Pictured below is an example of an on-farm settling basin. In this application, return flow from the farm is collected in a settling basin (foreground) before returning to the irrigation canal (background).

Range of Applications

Settling basins are used as a separation mechanism to eliminate rejected products (i.e. waste solids management strategies) of a specified size and quantity in various fields, such as aquaculture, mining, dairy, food processing, alcohol production and wine making. Regular draining and desilting of settling basins is required to maintain satisfactory performance.

Aquaculture

All materials not removed from the system during harvesting are categorized as wastes including uneaten feed, excreta, chemicals and therapeutics, dead and moribund fish, escaped fish and pathogens. Settling basins in the field are simple ponds dug downstream of the farm to optimally remove suspended solids effectively, produce clarified effluent, and accumulate and thicken sludge to minimal volume. If impairment occurs in any of these functions, this might have a great impact on pond performance, which could lead to damaging the effectiveness of the process.

Mining

Wastewater produced by mining industries contribute to the acidity, suspended material and dissolved heavy metal ions in the aquatic environment, causing environmental problems for biological life and discoloration of the receiving waters. The application of settling basins by the Coeur d'Alene mining district of northern Idaho, United States, globally known to produce lead, zinc and silver, to treat wastewater has greatly improved the quality of water discharge from mining operations.

Dairy Waste

By reducing flow velocity to limit solids being transported along with fast flowing liquid, separation can occur. Approximately 35% – 60% of the solids is removed from dilute liquid slurry, with 10 minutes detention time, with a common detention time of 30 to 60 minutes. Due to the inadequate consideration of critical design criteria, most settling basins built were oversized and had low efficiency.

Settling basins used in dairy production reduce the nutrient-loading on a vegetative filter strip from lot runoff, thus decreasing the required lagoon volume for a new facility. Moreover, settling basins are useful to remove unwanted solid materials, such as hay, straw and feathers from the waste stream before flowing to the lagoon, aids to reduce smell and avoid crust formation on the lagoon surface. A baffle may be used to retain the floating solids removed. There are two types of settling basins, based on the method of removing solids. With one type, the solids are removed mechanically (after the free water has drained away), usually with a front-end or skid-steer loader. The other type uses hydraulic (pump) removal of the solids. Typically, pumping is initiated when the basin is half full of solids and the remainder is water. Vigorous agitation is needed to mix the liquid and the solids, preferably by propeller-type agitators or pumps with agitation nozzles.

There are two types of settling basins, based on the method of removing solids. With one type, the solids are removed mechanically (after the free water has drained away), usually with a front-end or skid-steer loader. The depth of accumulated solids should not exceed 1.5 feet. The other type uses hydraulic (pump) removal of the solids. Typically, pumping is initiated when the basin is half full of solids and the remainder is water. Vigorous agitation is needed to mix the liquid and the solids, preferably by propeller-type agitators or pumps with agitation nozzles.

Settling basins may be either concrete or earthen structures. For concrete basins, a common recommendation is a minimum depth of 2 feet plus the depth required for solids storage. Figure shows a typical concrete settling basin. Earthen structures may be compact basins, settling terraces, settling diversion terraces or settling channels. Figure shows a typical earthen settling basin. Earthen basins to be cleaned with loaders are usually designed to be shallow (not more than 3 feet deep) and to cover a large area. Earthen settling basins should have a concrete entrance ramp and a concrete runway on the bottom to allow entry of equipment for solids removal. Figure shows a duplex concrete settling basin that allows one side to receive effluent while the other side is thoroughly dewatered and the solids removed.

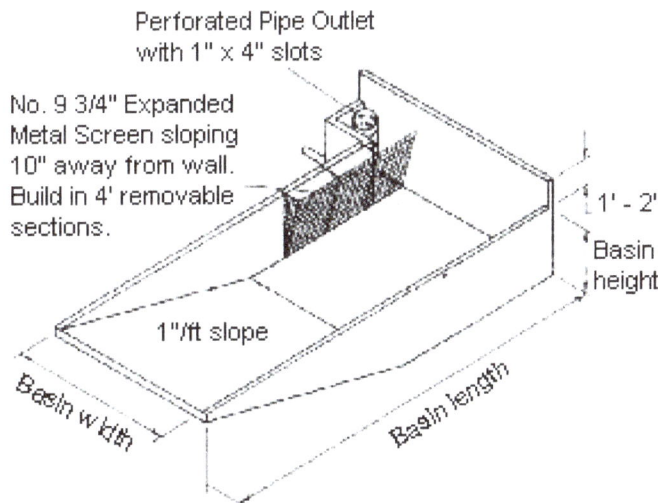

A typical concrete settling basin designed for mechanical removal of solids (from MWPS18.)

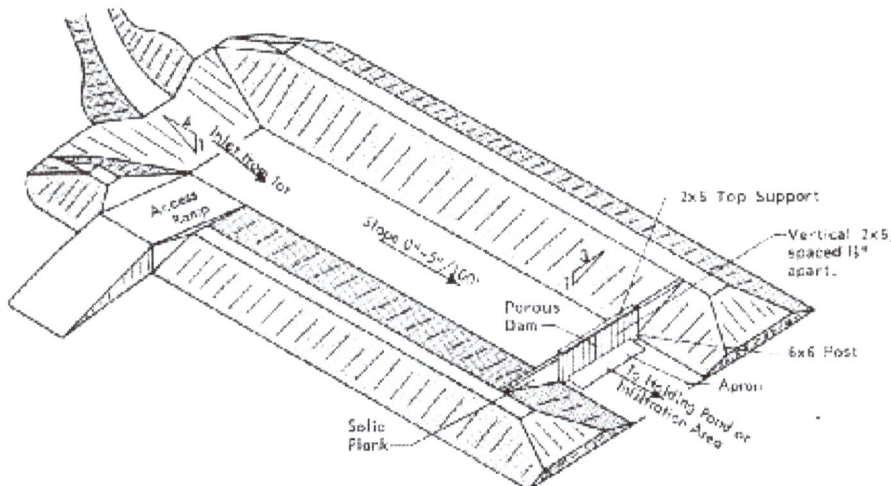

typical earthen settling basin designed for mechanical removal
of solids (from MWPS18)

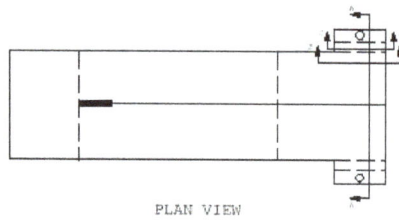

PLAN VIEW

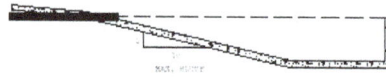

SIDE VIEW

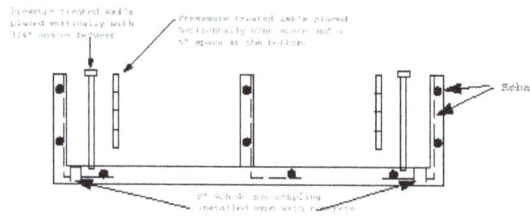

CROSS-SECTION A-A

CROSS-SECTION B-B

CROSS-SECTION C-C

Duplex concrete settling basin

A settling terrace, a settling diversion terrace or a settling channel is a wide, shallow, gently sloping, flat-bottomed waterway in which runoff solids settle due to low velocity. The channel is sometimes grassed to improve settling and reduce erosion. Grass may not survive in the channel and may make cleaning more difficult. Grass should be maintained on the sideslopes, if possible. Solids should be removed annually, or more often if required, to maintain capacity. Berm tops should be at least 2 feet wide to maintain the design height and at least 12 feet wide for vehicle traffic.

In Missouri's humid climate, inadequate drying of the solids and the channel may limit the usefulness of earthen settling terraces and channels.

Design of Concrete Settling Basins

An example will illustrate the design of a concrete settling basin. A blank form is included for your calculations.

Example

Design a concrete settling basin for a dirt lot, 100 feet by 200 feet, on a 6-percent slope. The solids will be pumped out at 6-month intervals. The basin length is to be four times the basin width. The location is in the northeast corner of Missouri. Design using peak runoff rate for a 1-year per 10-year storm from Table.

Table: Peak rates of runoff to be expected from watersheds in Missouri (Q)

Acres	One year per 10 (cubic feet per second per acre)	One year per 25[1] (cubic feet per second per acre)
0 to 1	5.5	0.2
1 to 2	4.6	6.0
2 to 3	4.2	5.5
3 to 4	3.0	5.2
4 to 20[2]	2.8	5.1

[1]One year per 10 is sufficient in most design situations. However, if a settling terrace is adjacent to a stream, use one year per 25.

[2]For areas greater than 20 acres, refer to MU publication G1518, Table 1.

Open Lot Area Draining into the Basin

Lot length 200 feet x lot width 100 feet = lot area 20,000 square feet

Inflow rate into settling basin

Q = peak runoff rate in cubic feet per second per acre
L = location factor
T = topographic factor
Q_T = inflow rate into settling basin, cubic feet per hour

QT = Q x L x T x square feet lot area x 0.0411 = cubic feet per hour

QT = 5.5 cubic feet per second per acre (Q) x 0.96 (L) x 0.92 (T) x 20,000 square feet x 0.0411 = 3,993 cubic feet per hour

If inflow arrives at settling basin via a sewer pipe, estimate QT from the size of the sewer pipe in Table.

QT = gpm (Table) x 8 = cubic feet per hour

If inflow arrives at settling basin by other means, explain (i.e., a dairy flush alley discharging into settling basin).

Estimate Inflow Rate

Q_T = estimated gpm x 8 = cubic feet per hour

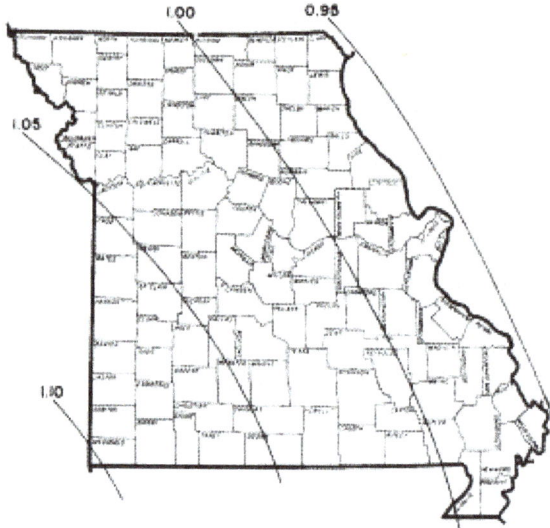

Location factors (L) from G1518, Estimating Peak
Rates of Runoff From Small Watersheds.

Table: Topographic factors for different land slopes.[1]

Average land slope	Topographic factor (T)
1 percent	0.65
2 percent	0.72
3 percent	0.78
4 percent	0.83
5 percent	0.88
6 percent	0.92
7 percent	0.96
8 percent	1.00
9 percent	1.04
10 percent	1.07
12 percent	1.14
14 percent	1.20

[1]Obtained from MU publication G1518, *Estimating Peak Rates of Runoff from Small Watersheds.*

Table: Diameter, slope and expected flow rates for sewer lines

Diameter	Slope	Expected flow rate
4 inches	0.02 foot per foot	115 GPM
6 inches	0.013 foot per foot	255 GPM
8 inches	0.009 foot per foot	460 GPM
10 inches	0.007 foot per foot	750 GPM

Surface Area (SA) of Settling Basin

SA (square feet)	=	Q_T (cubic feet per hour)/ 4 cubic feet per hour per square feet	=	3,993 cubic feet per hour/ 4 cubic feet per hour per square feet	=	998 square feet

Basin Dimensions

Design basin for length = 3 to 5 times basin width.

Basin width = $[SA/R]^{0.5}$

SA = basin surface area, square feet (from item 3)
R = length-width ratio = 4

width = [998 square feet (SA) ÷ 4 (R)]$^{0.5}$ = 15.8 feet

length = 16 feet (width) x 4 (R) = 64 feet

Basin Overflow

Provide a rectangular overflow weir at downstream end of basin. Weir height = 6 inches. Maximum weir length = width of settling basin. Minimum weir length, feet = Q_T/1,250. If a riser pipe is used as an overflow device instead of a rectangular weir, riser pipe diameter, inches = Q_T/274. Do not use a riser pipe smaller than 6 inches in diameter.

Minimum rectangular weir length = Q_T/1,250 = 3,993 /1,250 = 3.2 feet

Minimum riser pipe diameter = Q_T/274 = 3,993 /274 = 14.6 inches

Basin Depth

Use the following as a guide for volume of solids and calculate depth required for desired storage period.

Dirt lots = 2,800 cubic feet per acre-yr

Concrete lots and confinement buildings = 0.5 times manure production volume.

Table: Approximate daily manure production

Animal	Weight	Cubic feet per day (liquid + solid)
Dairy	1,000	1.32
Beef (a) Concentration ration (b) Corn silage (c) 25 percent haylage (d) 50 percent haylage	1,000 1,000 1,000 1,000	0.35 0.45 0.34 0.43
Swine, finish	1,000	1.10
Sow + litter	1,000	1.44
Gestation sow	1,000	0.55

Boar	1,000	0.55
Sheep	1,000	0.62
Poultry (a) Layers (b) Broilers	1,000 1,000 1,000	0.88 1.20
Horse	1,000	1.50

If solids are to be removed from basin by pumping, design basin to hold an equal volume of water above settled solids. Solids must be diluted and agitated for pumping. If solids are to be removed mechanically (i.e., front-end loader), provide concrete entrance to settling basin with at least 10:1 slope (20:1 slope preferred). Additional dewatering via a hardware cloth dam and/or a perforated riser pipe is desirable for mechanical removal of settled solids.

Indicate desired storage period, days = 182

(Lot acres = 100 feet x 200 feet/43,560 = 0.46 acres)

$$\text{Basin depth (dirt lot)} = \frac{2{,}800 \times 0.46 \text{ acres lot} \times 182 \text{ days storage} \times 2^*}{365 \text{ days per yr} \times 998 \text{ square feet surface area (item 3)}} = 1.3 \text{ feet}$$

*Multiply by 1 for mechanical removal or by 2 for pumping.

Basin depth, BD_c (concrete lot or confinement building)

$$BD_c = \frac{0.5 \times \underline{\quad} \text{ cubic feet per day manure prod (Table)} \times \underline{\quad} \text{ days storage} \times (1^* \text{ or } 2^*)}{\underline{\quad} \text{ square feet surface area (item 3)}} = \underline{\quad} \text{ feet}$$

*Multiply by 1 for mechanical removal or by 2 for pumping.

Note

When settling basin discharges into a lagoon, the size of the lagoon may be reduced as follows:
Design volume (with settling basin) = Design volume (without settling basin) x 0.5
Manure storage volume (with settling basin) = Manure storage volume (without settling basin) x 0.5

Minimum design storage period is 90 days when the lagoon design volume is reduced by 50 percent as noted above. Less storage may be used if the lagoon design volume is based on 100 percent loading.

Basin Outlets

Various types of basin outlets are used to drain liquids from the full depth of basins and allow the solids to dewater. The porous plank dam ahead of either a perforated or a slotted riser pipe are frequently used outlets. Manure tends to plug even large openings in outlets. Unplugging is required frequently. A hoe may be used to scrape solids off of openings. Also, a slanted expanded metal or quarry screen with 1-inch to 1.5-inch openings may be used around the outlet to increase the screening area and reduce clogging.

Porous Dams

Porous dams may be made of welded wire fabric, expanded metal mesh or spaced boards. Porous dams may be used to dewater settling basins or to remove large solids that tend to cause excessive clogging of the openings in perforated pipe outlets. Dams constructed with spaced boards usually have 0.75-inch spaces between the boards. The boards usually range from 2-by-6s to 2-by-12s. Expanded metal and welded wire fabric have even greater open areas. Due to the large open area in a porous dam, little design is required. As a general rule, the open area in a porous dam should be twice the area of the perforations in the riser pipe it precedes. As a practical matter, a porous dam 4 feet long or more, should suffice for the common sizes of outlet pipes. In some applications, there is no outlet pipe and the porous dam forms one wall of the settling/storage basin.

Perforated Pipe Outlets

Material for perforated pipe is usually PVC plastic, galvanized steel or concrete. Perforations can be 5/8-inch to 1-inch diameter holes or 1-inch by 4-inch slots. The outlet is sized to match the anticipated flow rates in order to assure adequate detention time. Flow rate is controlled by the amount of open area (slots or holes) in the pipe. Table gives opening requirements for perforated pipes.

Table: Riser pipe outlet design for settling basins (from MWPS18)

Based on
$Q = (C)(A)(2gh)^{0.5}$, where Q = flow rate in cfs; C = slot constant, assumed to be 0.61; A = open slot area in square feet; g = 32.174 feet per second^2; h = head on openings in feet. The pipe height is divided into 0.5-foot increments. The head on all the slots in the first 0.5-foot increment is assumed to be 0.25-foot. The head on subsequent 0.5-foot increments increase at 0.5-foot increments.

Open slot area per foot of pipe height, in ^{2} per foot	Head, feet							
	0.5	1.0	1.5	2.0	2.5	3.0	3.5	4.0
	cfs							
4	0.034	0.093	0.169	0.259	0.361	0.473	0.596	0.728
6	0.051	0.139	0.253	0.388	0.541	0.710	0.894	1.091
8	0.068	0.186	0.338	0.518	0.721	0.947	1.192	1.455
10	0.085	0.232	0.422	0.647	0.902	1.183	1.480	1.819
12	0.102	0.279	0.507	0.776	1.082	1.420	1.788	2.183
14	0.119	0.325	0.591	0.906	1.262	1.657	2.086	2.546
16	0.136	0.371	0.675	1.035	1.443	1.894	2.384	2.910
18	0.153	0.418	0.760	1.164	1.623	2.130	2.682	3.274
20	0.170	0.464	0.844	1.294	1.803	2.367	3.890	3.638
22	0.187	0.511	0.929	1.423	1.984	2.604	3.277	4.001
24	0.204	0.557	1.013	1.542	2.164	2.840	3.575	4.365
26	0.221	0.603	1.097	1.682	2.344	3.077	3.873	4.729
28	0.238	0.650	1.182	1.811	2.525	3.314	4.171	5.093
30	0.255	0.696	1.266	1.940	2.705	3.550	4.469	5.456
32	0.272	0.743	1.351	2.070	2.885	3.787	4.767	5.820
34	0.289	0.789	1.435	2.199	3.066	4.024	5.065	6.184

36	0.306	0.836	1.519	2.329	3.246	4.260	5.363	6.548
38	0.323	0.882	1.604	2.458	3.426	4.497	5.661	6.911
40	0.340	0.928	1.688	2.587	3.607	4.734	5.959	7.275

Example

Design a basin outlet to allow outflow to equal peak flow rate off the lot in Example 1 when the basin is full. The inflow rate in Example 1 is 3,993 cubic feet per hour and the depth is 1.3 feet.

- For a perforated pipe riser, determine the required open area per foot of pipe height from Table 5. Outflow = 3,993 cubic feet per hour = 1.1 cubic feet per second. By interpolation in Table 5, we find that with 1.3 feet of head the required opening area in the riser pipe is 32 in² per foot of pipe height.

- Size the outlet pipe from the data in Table 3. Outflow in gpm = 1.1 cubic feet per second x 450 gpm per cfs = 495 gpm. From Table 3, an 8-inch pipe at 0.009 slope will carry 460 gpm and a 10-inch pipe at 0.007 slope will carry 750 gpm. Depending on slope, use an 8-inch or 10-inch pipe (an 8-inch pipe will carry 1.1 cfs at slopes greater than 1 percent, Ref. Figure 4.5b in MWPS18).

Check on the time to withdraw the 9,167 cubic feet of runoff from the 25-year, 24-hour storm at 1.1 cfs.

Time in hours = 9,167 cubic feet ÷ (1.1 cubic feet per second x 3,600 second per hour) = 2.3 hours at maximum head.

Design of Earthen Settling Basins

To meet approval by the Missouri Department of Natural Resources, earthen basins must be built as follows. Berms shall have minimum slopes of 3:1. If solids are to be removed using mechanical equipment, a concrete pad shall be installed in the bottom of the basin and a concrete access ramp with a maximum slope of 10 percent shall be provided. If the settled solids are to be removed by pumping, the basin must be designed to contain an equal volume of water above the solids to allow for agitation and dilution of the solids. Access points for the mixing equipment must be indicated on the construction drawings.

Design of Settling Channels

A settling diversion terrace, settling terrace or a settling channel is a wide, shallow, gently sloping, flat-bottomed channel, in which suspended solids contained in runoff water are settled out. A settling channel may be either of earthen or concrete construction. The settling channel may be grassed to improve settling and reduce erosion. Runoff water from the channel is stored in a lagoon or storage pond. Wastes settled from the runoff are allowed to dry before removal with mechanical equipment, typically with a tractor and front-end loader. Usually, solids are removed from the channel once per year or when accumulated solids reduce the settling ability of the channel.

Sideslopes for settling channels usually range from 3:1 to 4:1, depending on soil properties. The bottom slope of the channel should be between 0.1 percent and 0.3 percent to maintain low velocities and rapid settling.

Sewage Treatment

Section of a wastewater treatment plant

Sewage treatment, or domestic wastewater treatment, is the process of removing contaminants from wastewater and household sewage, both runoff (effluents) and domestic. It includes physical, chemical and biological processes to remove physical, chemical and biological contaminants. Its objective is to produce a waste stream (or treated effluent) and a solid waste or sludge suitable for discharge or reuse back into the environment. This material is often inadvertently contaminated with many toxic organic and inorganic compounds.

Sewage is created by residences, institutions, hospitals and commercial and industrial establishments. It can be treated close to where it is created (in septic tanks, biofilters or aerobic treatment systems), or collected and transported via a network of pipes and pump stations to a municipal treatment plant. Sewage collection and treatment is typically subject to local, state and federal regulations and standards. Industrial sources of wastewater often require specialized treatment processes.

The sewage treatment involves three stages, called *primary*, *secondary*, and *tertiary treatment*. First, the solids are separated from the wastewater stream. Then, dissolved biological matter is progressively converted into a solid mass by using indigenous, water-borne micro-organisms. Finally, the biological solids are neutralized, then disposed of or re-used, and the treated water may be disinfected chemically or physically (for example by lagoons and micro-filtration). The final effluent can be discharged into a stream, river, bay, lagoon or wetland, or it can be used for the irrigation of a golf course, green way or park. If it is sufficiently clean, it can also be used for groundwater recharge or agricultural purposes.

Description

Raw influent (sewage) includes household waste liquid from toilets, baths, showers, kitchens, sinks, and so forth that is disposed of via sewers. In many areas, sewage also includes liquid waste from industry and commerce.

The separation and draining of household waste into greywater and blackwater is becoming more common in the developed world, with greywater being permitted to be used for watering plants or recycled for flushing toilets. A lot of sewage also includes some surface water from roofs or hard-standing areas. Municipal wastewater therefore includes residential, commercial, and industrial liquid waste discharges, and may include stormwater runoff. Sewage systems capable of handling stormwater are known as combined systems or combined sewers. Such systems are usually avoided since they complicate and thereby reduce the efficiency of sewage treatment plants owing to their seasonality. The variability in flow also leads to often larger than necessary, and subsequently more expensive, treatment facilities. In addition, heavy storms that contribute more flows than the treatment plant can handle may overwhelm the sewage treatment system, causing a spill or overflow (called a combined sewer overflow, or CSO, in the United States). It is preferable to have a separate storm drain system for stormwater in areas that are developed with sewer systems.

As rainfall runs over the surface of roofs and the ground, it may pick up various contaminants including soil particles and other sediment, heavy metals, organic compounds, animal waste, and oil and grease. Some jurisdictions require stormwater to receive some level of treatment before being discharged directly into waterways. Examples of treatment processes used for stormwater include sedimentation basins, wetlands, buried concrete vaults with various kinds of filters, and vortex separators (to remove coarse solids).

The site where the raw wastewater is processed before it is discharged back to the environment is called a wastewater treatment plant (WWTP). The order and types of mechanical, chemical and biological systems that comprise the wastewater treatment plant are typically the same for most developed countries:

- Mechanical treatment

  - Influx (Influent)

  - Removal of large objects

  - Removal of sand and grit

  - Pre-precipitation

- Biological treatment

  - Oxidation bed (oxidizing bed) or aeration system

  - Post precipitation

- Chemical treatment this step is usually combined with settling and other processes to remove solids, such as filtration. The combination is referred to in the United States as physical chemical treatment.

Primary treatment removes the materials that can be easily collected from the raw wastewater and disposed of. The typical materials that are removed during primary treatment include fats, oils, and greases (also referred to as FOG), sand, gravels and rocks (also referred to as grit), larger settleable solids and floating materials (such as rags and flushed feminine hygiene products). This step is done entirely with machinery.

Process Flow Diagram for a Typical Large-Scale Treatment Plant

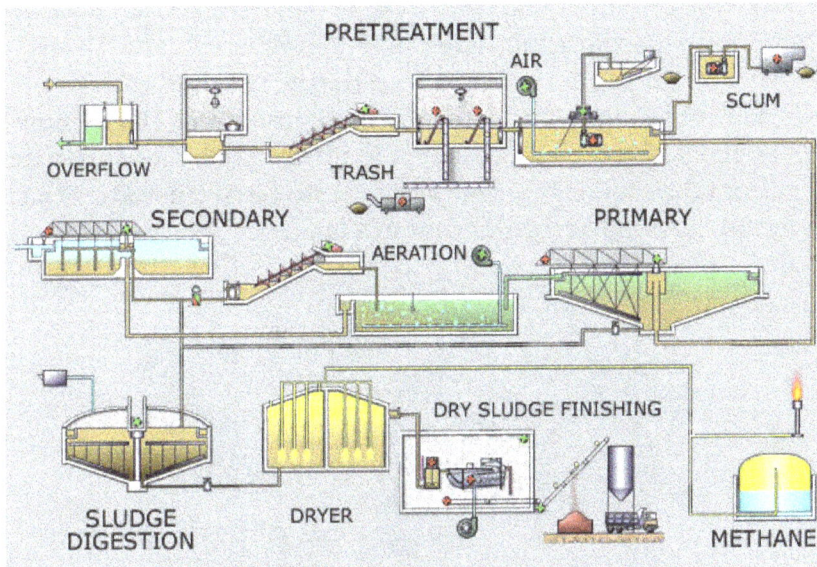

Process Flow Diagram for a Typical Treatment Plant via Subsurface Flow Constructed Wetlands (SFCW)

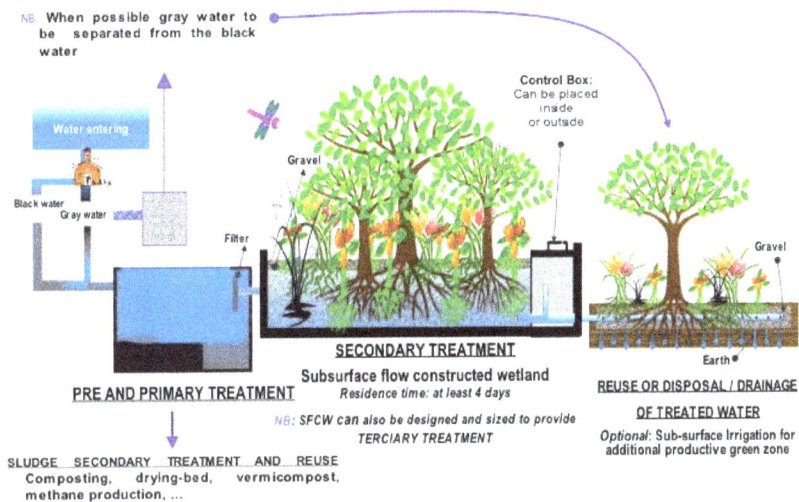

Primary treatment

Removal of Large Objects from Influent Sewage

In primary treatment, the influent sewage water is strained to remove all large objects that are deposited in the sewer system, such as rags, sticks, tampons, cans, fruit, etc. This is most commonly done with a manual or automated mechanically raked bar screen. The raking action of a mechanical bar screen is typically paced according to the accumulation on the bar screens and/or flow rate. The bar screen is used because large solids can damage or clog the equipment used later in the sewage treatment plant. The solids are collected in a dumpster and later disposed in a landfill.

Primary treatment also typically includes a sand or grit channel or chamber where the velocity of the incoming wastewater is carefully controlled to allow sand grit and stones to settle, while keeping the majority of the suspended organic material in the water column. This equipment is called a degritter or sand catcher. Sand, grit, and stones need to be removed early in the process to avoid damage to pumps and other equipment in the remaining treatment stages. Sometimes there is a sand washer (grit classifier) followed by a conveyor that transports the sand to a container for disposal. The contents from the sand catcher may be fed into the incinerator in a sludge processing plant, but in many cases, the sand and grit is sent to a landfill.

Empty sedimentation tank at the treatment plant in Merchtem

Sedimentation

Many plants have a sedimentation stage where the sewage is allowed to pass slowly through large tanks, commonly called "primary clarifiers" or "primary sedimentation tanks." The tanks are large enough that sludge can settle and floating material such as grease and oils can rise to the surface and be skimmed off. The main purpose of the primary clarification stage is to produce both a generally homogeneous liquid capable of being treated biologically and a sludge that can be separately treated or processed. Primary settling tanks are usually equipped with mechanically driven scrapers that continually drive the collected sludge towards a hopper in the base of the tank from where it can be pumped to further sludge treatment stages.

Secondary Treatment

Secondary treatment is designed to substantially degrade the biological content of the sewage such as are derived from human waste, food waste, soaps and detergent. The majority of municipal plants treat the settled sewage liquor using aerobic biological processes. For this to be effective, the biota require both oxygen and a substrate on which to live. There are a number of ways in which this is done. In all these methods, the bacteria and protozoa consume biodegradable soluble organic contaminants (e.g. sugars, fats, organic short-chain carbon molecules, etc.) and bind much of the less soluble fractions into floc. Secondary treatment systems are classified as fixed film or suspended growth. Fixed-film treatment process including trickling filter and rotating biological contactors where the biomass grows on media and the sewage passes over its surface. In suspended growth systems—such as activated sludge—the biomass is well mixed with the sewage and can be operated in a smaller space than fixed-film systems that treat the same amount of water.

However, fixed-film systems are more able to cope with drastic changes in the amount of biological material and can provide higher removal rates for organic material and suspended solids than suspended growth systems.

Roughing filters are intended to treat particularly strong or variable organic loads, typically industrial, to allow them to then be treated by conventional secondary treatment processes. Characteristics include typically tall, circular filters filled with open synthetic filter media to which wastewater is applied at a relatively high rate. They are designed to allow high hydraulic loading and a high flow-through of air. On larger installations, air is forced through the media using blowers. The resultant wastewater is usually within the normal range for conventional treatment processes.

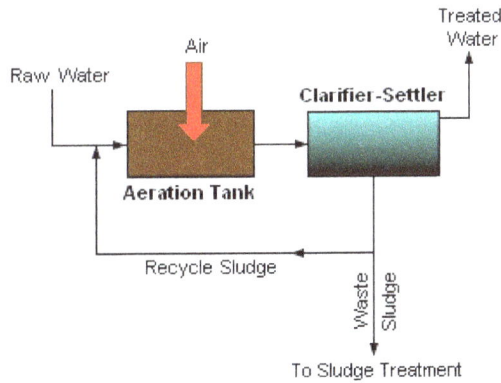

A generalized, schematic diagram of an activated sludge process.

Activated Sludge

In general, activated sludge plants encompass a variety of mechanisms and processes that use dissolved oxygen to promote the growth of biological floc that substantially removes organic material.

The process traps particulate material and can, under ideal conditions, convert ammonia to nitrite and nitrate and ultimately to nitrogen gas.

Surface-aerated Basins

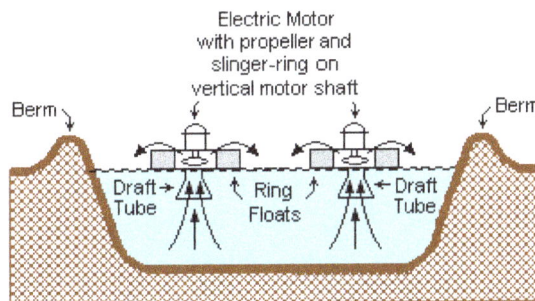

A TYPICAL SURFACE – AERATED BASIN

Note: The ring floats are tethered to posts on the berms.

A Typical Surface-Aerated Basin (using motor-driven floating aerators)

Most biological oxidation processes for treating industrial wastewaters have in common the use of oxygen (or air) and microbial action. Surface-aerated basins achieve 80 to 90 percent removal of Biochemical Oxygen Demand with retention times of 1 to 10 days. The basins may range in depth from 1.5 to 5.0 meters and use motor-driven aerators floating on the surface of the wastewater.

In an aerated basin system, the aerators provide two functions: they transfer air into the basins required by the biological oxidation reactions, and they provide the mixing required for dispersing the air and for contacting the reactants (that is, oxygen, wastewater and microbes). Typically, the floating surface aerators are rated to deliver the amount of air equivalent to 1.8 to 2.7 kg O_2/kW•h. However, they do not provide as good mixing as is normally achieved in activated sludge systems and therefore aerated basins do not achieve the same performance level as activated sludge units.

Biological oxidation processes are sensitive to temperature and, between 0 °C and 40 °C, the rate of biological reactions increase with temperature. Most surface aerated vessels operate at between 4 °C and 32 °C.

Fluidized Bed Reactors

The carbon absorption following biological treatment is particularly effective in reducing both the BOD and COD to low levels. A fluidized bed reactor is a combination of the most common stirred tank packed bed, continuous flow reactors. It is very important to chemical engineering because of its excellent heat and mass transfer characteristics. In a fluidized bed reactor, the substrate is passed upward through the immobilized enzyme bed at a high velocity to lift the particles. However the velocity must not be so high that the enzymes are swept away from the reactor entirely. This causes low mixing; these type of reactors are highly suitable for the exothermic reactions. It is most often applied in immobilized enzyme catalysis

Filter Beds (Oxidizing Beds)

In older plants and plants receiving more variable loads, trickling filter beds are used where the settled sewage liquor is spread onto the surface of a deep bed made up of coke (carbonized coal), limestone chips or specially fabricated plastic media. Such media must have high surface areas to support the biofilms that form. The liquor is distributed through perforated rotating arms radiating from a central pivot. The distributed liquor trickles through this bed and is collected in drains at the base. These drains also provide a source of air which percolates up through the bed, keeping it aerobic. Biological films of bacteria, protozoa and fungi form on the media's surfaces and eat or otherwise reduce the organic content. This biofilm is grazed by insect larvae and worms which help maintain an optimal thickness. Overloading of beds increases the thickness of the film leading to clogging of the filter media and ponding on the surface.

Biological Aerated Filters

Biological Aerated (or Anoxic) Filter (BAF) or Biofilters combine filtration with biological carbon reduction, nitrification or denitrification. BAF usually includes a reactor filled with a filter media. The media is either in suspension or supported by a gravel layer at the foot of the filter. The dual purpose of this media is to support highly active biomass that is attached to it and to filter suspended solids. Carbon reduction and ammonia conversion occurs in aerobic mode and sometime achieved in a single reactor while nitrate conversion occurs in anoxic mode. BAF is operated either in upflow or downflow configuration depending on design specified by manufacturer.

Secondary Sedimentation tank at a rural treatment plant

Membrane Bioreactors

Membrane bioreactors (MBR) combines activated sludge treatment with a membrane liquid-solid separation process. The membrane component uses low pressure microfiltration or ultra filtration membranes and eliminates the need for clarification and tertiary filtration. The membranes are typically immersed in the aeration tank (however, some applications utilize a separate membrane tank). One of the key benefits of a membrane bioreactor system is that it effectively overcomes the limitations associated with poor settling of sludge in conventional activated sludge (CAS) processes. The technology permits bioreactor operation with considerably higher mixed liquor suspended solids (MLSS) concentration than CAS systems, which are limited by sludge settling. The process is typically operated at MLSS in the range of 8,000–12,000 mg/L, while CAS are operated in the range of 2,000–3,000 mg/L. The elevated biomass concentration in the membrane bioreactor process allows for very effective removal of both soluble and particulate biodegradable materials at higher loading rates. Thus increased Sludge Retention Times (SRTs)—usually exceeding 15 days— ensure complete nitrification even in extremely cold weather.

The cost of building and operating a MBR is usually higher than conventional wastewater treatment, however, as the technology has become increasingly popular and has gained wider acceptance throughout the industry, the life-cycle costs have been steadily decreasing. The small footprint of MBR systems, and the high quality effluent produced, makes them particularly useful for water reuse applications.

Secondary Sedimentation

The final step in the secondary treatment stage is to settle out the biological floc or filter material and produce sewage water containing very low levels of organic material and suspended matter.

Rotating Biological Contactors

Rotating biological contactors (RBCs) are mechanical secondary treatment systems, which are robust and capable of withstanding surges in organic load. RBCs were first installed in Germany in 1960 and have since been developed and refined into a reliable operating unit. The rotating disks support the growth of bacteria and micro-organisms present in the sewage, which breakdown and stabilize organic pollutants. To be successful, micro-organisms need both oxygen to live and food to grow. Oxygen is obtained from the atmosphere as the disks rotate. As the micro-organisms grow, they build up on the media until they are sloughed off due to shear forces provided by the rotating discs in the sewage. Effluent from the RBC is then passed through final clarifiers where the micro-organisms in suspension settle as a sludge. The sludge is withdrawn from the clarifier for further treatment.

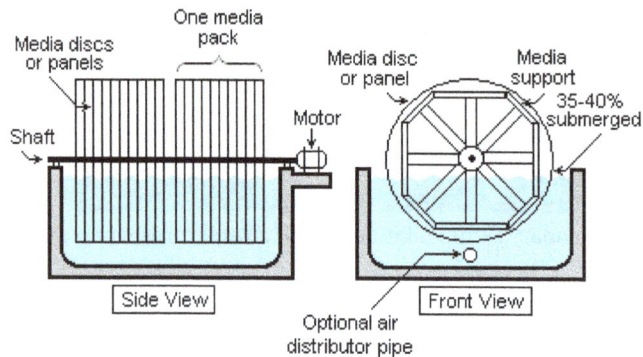

Schematic diagram of a typical rotating biological contactor (RBC). The treated effluent clarifier/settler is not included in the diagram.

A functionally similar biological filtering system has become popular as part of home aquarium filtration and purification. The aquarium water is drawn up out of the tank and then cascaded over a freely spinning corrugated fiber-mesh wheel before passing through a media filter and back into the aquarium. The spinning mesh wheel develops a biofilm coating of microorganisms that feed on the suspended wastes in the aquarium water and are also exposed to the atmosphere as the wheel rotates. This is especially good at removing waste urea and ammonia urinated into the aquarium water by the fish and other animals.

Tertiary Treatment

The purpose of tertiary treatment is to provide a final treatment stage to raise the effluent quality before it is discharged to the receiving environment (sea, river, lake, ground, etc.). More than one tertiary treatment process may be used at any treatment plant. If disinfection is practiced, it is always the final process. It is also called "effluent polishing."

Filtration

Sand filtration removes much of the residual suspended matter. Filtration over activated carbon removes residual toxins.

Lagooning

Lagooning provides settlement and further biological improvement through storage in large man-made ponds or lagoons. These lagoons are highly aerobic and colonization by native macrophytes, especially reeds, is often encouraged. Small filter feeding invertebrates such as Daphnia and species of Rotifera greatly assist in treatment by removing fine particulates.

A sewage treatment plant and lagoon in Everett, Washington.

Constructed Wetlands

Constructed wetlands include engineered reedbeds and a range of similar methodologies, all of which provide a high degree of aerobic biological improvement and can often be used instead of secondary treatment for small communities, also see phytoremediation. One example is a small reedbed used to clean the drainage from the elephants' enclosure at Chester Zoo in England.

Nutrient Removal

Wastewater may contain high levels of the nutrients nitrogen and phosphorus. Excessive release to the environment can lead to a build up of nutrients, called eutrophication, which can in turn encourage the overgrowth of weeds, algae, and cyanobacteria (blue-green algae). This may cause an algal bloom, a rapid growth in the population of algae. The algae numbers are unsustainable and eventually most of them die. The decomposition of the algae by bacteria uses up so much of oxygen in the water that most or all of the animals die, which creates more organic matter for the bacteria to decompose. In addition to causing deoxygenation, some algal species produce toxins that contaminate drinking water supplies. Different treatment processes are required to remove nitrogen and phosphorus.

Nitrogen Removal

The removal of nitrogen is effected through the biological oxidation of nitrogen from ammonia (nitrification) to nitrate, followed by denitrification, the reduction of nitrate to nitrogen gas. Nitrogen gas is released to the atmosphere and thus removed from the water.

Nitrification itself is a two-step aerobic process, each step facilitated by a different type of bacteria. The oxidation of ammonia (NH_3) to nitrite (NO_2^-) is most often facilitated by *Nitrosomonas* spp. (nitroso referring to the formation of a nitroso functional group). Nitrite oxidation to nitrate (NO_3^-), though traditionally believed to be facilitated by *Nitrobacter* spp. (nitro referring the formation of a nitro functional group), is now known to be facilitated in the environment almost exclusively by *Nitrospira* spp.

Denitrification requires anoxic conditions to encourage the appropriate biological communities to form. It is facilitated by a wide diversity of bacteria. Sand filters, lagooning and reed beds can all be used to reduce nitrogen, but the activated sludge process (if designed well) can do the job the most easily. Since denitrification is the reduction of nitrate to dinitrogen gas, an electron donor is needed. This can be, depending on the wastewater, organic matter (from faeces), sulfide, or an added donor like methanol.

Sometimes the conversion of toxic ammonia to nitrate alone is referred to as tertiary treatment.

Phosphorus Removal

Phosphorus removal is important as it is a limiting nutrient for algae growth in many fresh water systems (for negative effects of algae see Nutrient removal). It is also particularly important for water reuse systems where high phosphorus concentrations may lead to fouling of downstream equipment such as reverse osmosis.

Phosphorus can be removed biologically in a process called enhanced biological phosphorus removal. In this process, specific bacteria, called polyphosphate accumulating organisms (PAOs), are selectively enriched and accumulate large quantities of phosphorus within their cells (up to 20 percent of their mass). When the biomass enriched in these bacteria is separated from the treated water, these biosolids have a high fertilizer value.

Phosphorus removal can also be achieved by chemical precipitation, usually with salts of iron (e.g. ferric chloride), aluminum (e.g. alum), or lime. This may lead to excessive sludge productions as hydroxides precipitates and the added chemicals can be expensive. Despite this, chemical phosphorus removal requires significantly smaller equipment footprint than biological removal, is easier to operate and is often more reliable than biological phosphorus removal.

Once removed, phosphorus, in the form of a phosphate rich sludge, may be land filled or, if in suitable condition, resold for use in fertilizer.

Disinfection

The purpose of disinfection in the treatment of wastewater is to substantially reduce the number of microorganisms in the water to be discharged back into the environment. The effectiveness of disinfection depends on the quality of the water being treated (e.g., cloudiness, pH, etc.), the type of disinfection being used, the disinfectant dosage (concentration and time), and other environmental variables. Cloudy water will be treated less successfully since solid matter can shield organisms, especially from ultraviolet light or if contact times are low. Generally, short contact times, low doses and high flows all militate against effective disinfection. Common methods of disinfection include ozone, chlorine, or ultraviolet light. Chloramine, which is used for drinking water, is not used in wastewater treatment because of its persistence.

Chlorination remains the most common form of wastewater disinfection in North America due to its low cost and long-term history of effectiveness. One disadvantage is that chlorination of residual organic material can generate chlorinated-organic compounds that may be carcinogenic or harmful to the environment. Residual chlorine or chloramines may also be capable of chlorinating organic material in the natural aquatic environment. Further, because residual chlorine is toxic to aquatic species, the treated effluent must also be chemically dechlorinated, adding to the complexity and cost of treatment.

Ultraviolet (UV) light can be used instead of chlorine, iodine, or other chemicals. Because no chemicals are used, the treated water has no adverse effect on organisms that later consume it, as may be the case with other methods. UV radiation causes damage to the genetic structure of bacteria, viruses, and other pathogens, making them incapable of reproduction. The key disadvantages of UV disinfection are the need for frequent lamp maintenance and replacement and the need for a highly treated effluent to ensure that the target microorganisms are not shielded from the UV radiation (i.e., any solids present in the treated effluent may protect microorganisms from the UV light). In the United Kingdom, light is becoming the most common means of disinfection because of the concerns about the impacts of chlorine in chlorinating residual organics in the wastewater and in chlorinating organics in the receiving water. Edmonton, Alberta, Canada also uses UV light for its water treatment.

Ozone O_3 is generated by passing oxygen O_2 through a high voltage potential resulting in a third oxygen atom becoming attached and forming O_3. Ozone is very unstable and reactive and oxidizes most organic material it comes in contact with, thereby destroying many pathogenic microorganisms. Ozone is considered to be safer than chlorine because, unlike chlorine which has to be stored on site (highly poisonous in the event of an accidental release), ozone is generated onsite as needed. Ozonation also produces fewer disinfection by-products than chlorination. A disadvantage of ozone disinfection is the high cost of the ozone generation equipment and the requirements for special operators.

Package Plants and Batch Reactors

In order to use less space, treat difficult waste, deal with intermittent flow or achieve higher environmental standards, a number of designs of hybrid treatment plants have been produced. Such plants often combine all or at least two stages of the three main treatment stages into one combined stage. In the UK, where a large number of sewage treatment plants serve small populations, package plants are a viable alternative to building discrete structures for each process stage.

One type of system that combines secondary treatment and settlement is the sequencing batch reactor (SBR). Typically, activated sludge is mixed with raw incoming sewage and mixed and aerated. The resultant mixture is then allowed to settle producing a high quality effluent. The settled sludge is run off and re-aerated before a proportion is returned to the head of the works. SBR plants are now being deployed in many parts of the world including North Liberty, Iowa, and Llanasa, North Wales.

The disadvantage of such processes is that precise control of timing, mixing and aeration is required. This precision is usually achieved by computer controls linked to many sensors in the plant. Such a complex, fragile system is unsuited to places where such controls may be unreliable, or poorly maintained, or where the power supply may be intermittent.

Package plants may be referred to as *high charged* or *low charged*. This refers to the way the biological load is processed. In high charged systems, the biological stage is presented with a high organic load and the combined floc and organic material is then oxygenated for a few hours before being charged again with a new load. In the low charged system the biological stage contains a low organic load and is combined with flocculate for a relatively long time.

Sludge Treatment and Disposal

The sludges accumulated in a wastewater treatment process must be treated and disposed of in a safe and effective manner. The purpose of digestion is to reduce the amount of organic matter and the number of disease-causing microorganisms present in the solids. The most common treatment options include anaerobic digestion, aerobic digestion, and composting.

Choice of a wastewater solid treatment method depends on the amount of solids generated and other site-specific conditions. However, in general, composting is most often applied to smaller-scale applications followed by aerobic digestion and then lastly anaerobic digestion for the larger-scale municipal applications.

Anaerobic Digestion

Anaerobic digestion is a bacterial process that is carried out in the absence of oxygen. The process can either be *thermophilic* digestion, in which sludge is fermented in tanks at a temperature of 55°C, or *mesophilic*, at a temperature of around 36°C. Though allowing shorter retention time (and thus smaller tanks), thermophilic digestion is more expensive in terms of energy consumption for heating the sludge.

One major feature of anaerobic digestion is the production of biogas, which can be used in generators for electricity production and/or in boilers for heating purposes.

Aerobic Digestion

Aerobic digestion is a bacterial process occurring in the presence of oxygen. Under aerobic conditions, bacteria rapidly consume organic matter and convert it into carbon dioxide. The operating costs used to be characteristically much greater for aerobic digestion because of the energy used by the blowers, pumps, and motors needed to add oxygen to the process. However, recent technological advances include non-electric aerated filter systems that use natural air currents for the aeration instead of electrically operated machinery. Aerobic digestion can also be achieved by using diffuser systems or jet aerators to oxidize the sludge.

Composting

Composting is also an aerobic process that involves mixing the sludge with sources of carbon such as sawdust, straw or wood chips. In the presence of oxygen, bacteria digest both the wastewater solids and the added carbon source and, in doing so, produce a large amount of heat.

Sludge Disposal

When a liquid sludge is produced, further treatment may be required to make it suitable for final disposal. Typically, sludges are thickened (dewatered) to reduce the volumes transported off-site for disposal. There is no process which completely eliminates the need to dispose of biosolids. There is, however, an additional step some cities are taking to superheat the wastewater sludge and convert it into small pelletized granules that are high in nitrogen and other organic materials. In New York City, for example, several sewage treatment plants have dewatering facilities that use large centrifuges along with the addition of chemicals such as polymer to further remove liquid from the sludge. The removed fluid, called centrate, is typically reintroduced into the wastewater process. The product which is left is called "cake" and that is picked up by companies which turn it into fertilizer pellets. This product is then sold to local farmers and turf farms as a soil amendment or fertilizer, reducing the amount of space required to dispose of sludge in landfills.

Treatment in the Receiving Environment

The outlet of a wastewater treating plant flows into a small river.

Many processes in a wastewater treatment plant are designed to mimic the natural treatment processes that occur in the environment, whether that environment is a natural water body or the ground. If not overloaded, bacteria in the environment will consume organic contaminants, although this will reduce the levels of oxygen in the water and may significantly change the overall ecology of the receiving water. Native bacterial populations feed on the organic contaminants, and the numbers of disease-causing microorganisms are reduced by natural environmental conditions such as predation exposure to ultraviolet radiation, for example. Consequently, in cases where the receiving environment provides a high level of dilution, a high degree of wastewater treatment may not be required. However, recent evidence has demonstrated that very low levels of certain contaminants in wastewater, including hormones (from animal husbandry and residue from human hormonal contraception methods) and synthetic materials such as phthalates that mimic hormones in their action, can have an unpredictable adverse impact on the natural biota and potentially on humans if the water is re-used for drinking water. In the United States and EU, uncontrolled discharges of wastewater to the environment are not permitted under law, and strict water quality requirements are to be met. A significant threat in the coming decades will be the increasing uncontrolled discharges of wastewater within rapidly developing countries.

Sewage Treatment in Developing Countries

There are few reliable figures on the share of the wastewater collected in sewers that is being treated in the world. In many developing countries the bulk of domestic and industrial wastewater is discharged without any treatment or after primary treatment only. In Latin America about 15 percent of collected wastewater passes through treatment plants (with varying levels of actual treatment). In Venezuela, a below average country in South America with respect to wastewater treatment, 97 percent of the country's sewage is discharged raw into the environment.

In a relatively developed Middle Eastern country such as Iran, Tehran's majority of population has totally untreated sewage injected to the city's groundwater. Israel has also aggressively pursued the use of treated sewer water for irrigation. In 2008, agriculture in Israel consumed 500 million cubic meters of potable water and an equal amount of treated sewer water. The country plans to provide a further 200 million cubic meters of recycled sewer water and build more desalination plants to supply even more water.

Most of sub-saharan Africa is without wastewater treatment.

Water utilities in developing countries are chronically underfunded because of low water tariffs, the nonexistence of sanitation tariffs in many cases, low billing efficiency (i.e. many users that are billed do not pay) and poor operational efficiency (i.e. there are overly high levels of staff, there are high physical losses, and many users have illegal connections and are thus not being billed). In addition, wastewater treatment typically is the process within the utility that receives the least attention, partly because enforcement of environmental standards is poor. As a result of all these factors, operation and maintenance of many wastewater treatment plants is poor. This is evidenced by the frequent breakdown of equipment, shutdown of electrically operated equipment due to power outages or to reduce costs, and sedimentation due to lack of sludge removal.

Developing countries as diverse as Egypt, Algeria, China or Colombia have invested substantial sums in wastewater treatment without achieving a significant impact in terms of environmental improvement. Even if wastewater treatment plants are properly operating, it can be argued that the environmental impact is limited in cases where the assimilative capacity of the receiving waters (ocean with strong currents or large rivers) is high, as it is often the case.

Enhanced Biological Phosphorus Removal

Enhanced Biological Phosphorus Removal is a process in which the biology within the activated sludge system is manipulated so that microorganisms take more phosphorus into their cells than they normally would.

It should be noted that all activated sludge systems involve biological phosphorus removal; phosphorus is an essential element for life. The key to EBPR is getting microorganisms to take in more phosphorus than they need.

This is also known as luxury uptake.

Phosphorous Removal from Wastewater

Controlling phosphorous discharged from municipal and industrial wastewater treatment plants is a key factor in preventing eutrophication of surface waters. Phosphorous is one of the major nutrients contributing in the increased eutrophication of lakes and natural waters. Its presence causes many water quality problems including increased purification costs, decreased recreational and conservation value of an impoundments, loss of livestock and the possible lethal effect of algal toxins on drinking water.

Phosphate removal is currently achieved largely by chemical precipitation, which is expensive and causes an increase of sludge volume by up to 40%. An alternative is the biological phosphate removal (BPR).

Phosphorous in Wastewater

Municipal wastewaters may contain from 5 to 20 mg/l of total phosphorous, of which 1-5 mg/l is organic and the rest in inorganic. The individual contribution tend to increase, because phosphorous is one of the main constituent of synthetic detergents. The individual phosphorous contribution varies between 0.65 and 4.80 g/inhabitant per day with an average of about 2.18 g. The usual forms of phosphorous found in aqueous solutions include:

- Orthophosphates: available for biological metabolism without further breakdown.

- Polyphosphates: molecules with 2 or more phosphorous atoms, oxygen and in some cases hydrogen atoms combine in a complex molecule. Usually polyphosphates undergo hydrolysis and revert to the orthophosphate forms. This process is usually quite slow.

Normally secondary treatment can only remove 1-2 mg/l, so a large excess of phosphorous is discharged in the final effluent, causing eutrophication in surface waters. New legislation requires a maximum concentration of P discharges into sensitive water of 2 mg/l.

Phosphorous Removal Processes

The removal of phosphorous from wastewater involves the incorporation of phosphate into TSS and the subsequent removal from these solids. Phosphorous can be incorporated into either biological solids (e.g. micro organisms) or chemical precipitates.

Phosphate Precipitation

Chemical precipitation is used to remove the inorganic forms of phosphate by the addition of a coagulant and a mixing of wastewater and coagulant. The multivalent metal ions most commonly used are calcium, aluminium and iron.

Calcium:

it is usually added in the form of lime $Ca(OH)_2$. It reacts with the natural alkalinity in the wastewater to produce calcium carbonate, which is primarily responsible for enhancing SS removal.

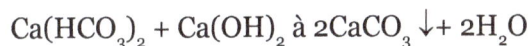

$$Ca(HCO_3)_2 + Ca(OH)_2 \rightarrow 2CaCO_3 \downarrow + 2H_2O$$

As the pH value of the wastewater increases beyond about 10, excess calcium ions will then react with the phosphate, to precipitate in hydroxylapatite:

$$10\ Ca^{2+} + 6\ PO_4^{3-} + 2\ OH^- \leftrightarrow Ca_{10}(PO_4)*6(OH)_2 \downarrow$$

Because the reaction is between the lime and the alkalinity of the wastewater, the quantity required will be, in general, independent of the amount of phosphate present. It will depend primarily on the alkalinity of the wastewater. The lime dose required can be approximated at 1.5 times the alkalinity as $CaCO_3$. Neutralisation may be required to reduce pH before subsequent treatment or disposal. Recarbonation with carbon dioxide (CO_2) is used to lower the pH value.

Aluminium and Iron:

Alum or hydrated aluminium sulphate is widely used precipitating phosphates and aluminium phosphates ($AlPO_4$). The basic reaction is:

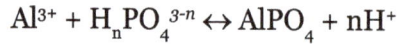

$$Al^{3+} + H_nPO_4^{3-n} \leftrightarrow AlPO_4 + nH^+$$

This reaction is deceptively simple and must be considered in light of the many competing reactions and their associated equilibrium constants and the effects of alkalinity, pH, trace elements found in wastewater. The dosage rate required is a function of the phosphorous removal required. The efficiency of coagulation falls as the concentration of phosphorous decreases. In practice, an 80-90% removal rate is achieved at coagulant dosage rates between 50 and 200 mg/l. Dosages are generally established on the basis of bench-scale tests and occasionally by full-scale tests, especially if polymers are used. Aluminium coagulants can adversely affect the microbial population in activated sludge, especially protozoa and rotifers, at dosage rates higher than 150 mg/l. However this does not affect much either BOD or TSS removal, as the clarification function of protozoa and rotifers is largely compensated by the enhanced removal of SS by chemical precipitation.

Ferric chloride or sulphate and ferrous sulphate also know as copperas, are all widely used for phosphorous removal, although the actual reactions are not fully understood. The basic reaction is:

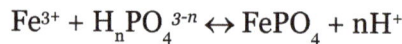

$$Fe^{3+} + H_nPO_4^{3-n} \leftrightarrow FePO_4 + nH^+$$

Ferric ions combine to form ferric phosphate. They react slowly with the natural alkalinity and so a coagulant aid, such as lime, is normally add to raise the pH in order to enhance the coagulation.

Strategies

The main phosphate removal processes are:

1.   Treatment of raw/primary wastewater;

2.   Treatment of final effluent of biological plants (postprecipitation);

3.   Treatment contemporary to the secondary biologic reaction (co-precipitation).

The first process is included in the general category of chemical precipitation processes. Phosphorous is removed with 90% efficiency and the final P concentration is lower than 0.5 mg/l. The chemical dosage for P removal is the same as the dosage needed for BOD and SS removal, which uses the main part of these chemicals. As mentioned above lime consumption is dependent on the alkalinity of the wastewater: only 10% of the lime fed is used in the phosphorous removal reaction. The remaining amount reacts with water alkalinity, with softening. To determine the lime quantity needed it is possible to use diagrams: i.e. the lime used to reach ph 11 is 2-2.5 times water alkalinity.

The postprecipitation is a standard treatment of a secondary effluent, usually using only metallic reagents. It is the process that gives the highest efficiency in phosphorous removal. Efficiency can reach 95%, and P concentration in the effluent can be lower than 0.5 mg/l. Postprecipitation gives also a good removal of the SS that escape the final sedimentation of the secondary process.

Its advantage is also to guarantee purification efficiency at a certain extent even if the biological process is not efficient for some reason. The chemical action is stronger, since the previous biologic treatment transforms part of the organic phosphates in orthophosphates. Disadvantages are high costs for the treatment plant (big ponds and mixing devices) and sometimes a too dilute effluent. Using ferric salts there is also the risk of having some iron in the effluent, with residual coloration. The metallic ions dosage is about 1.5-2.5 ions for every phosphorus ion (on average about 10-30 g/mc of water).

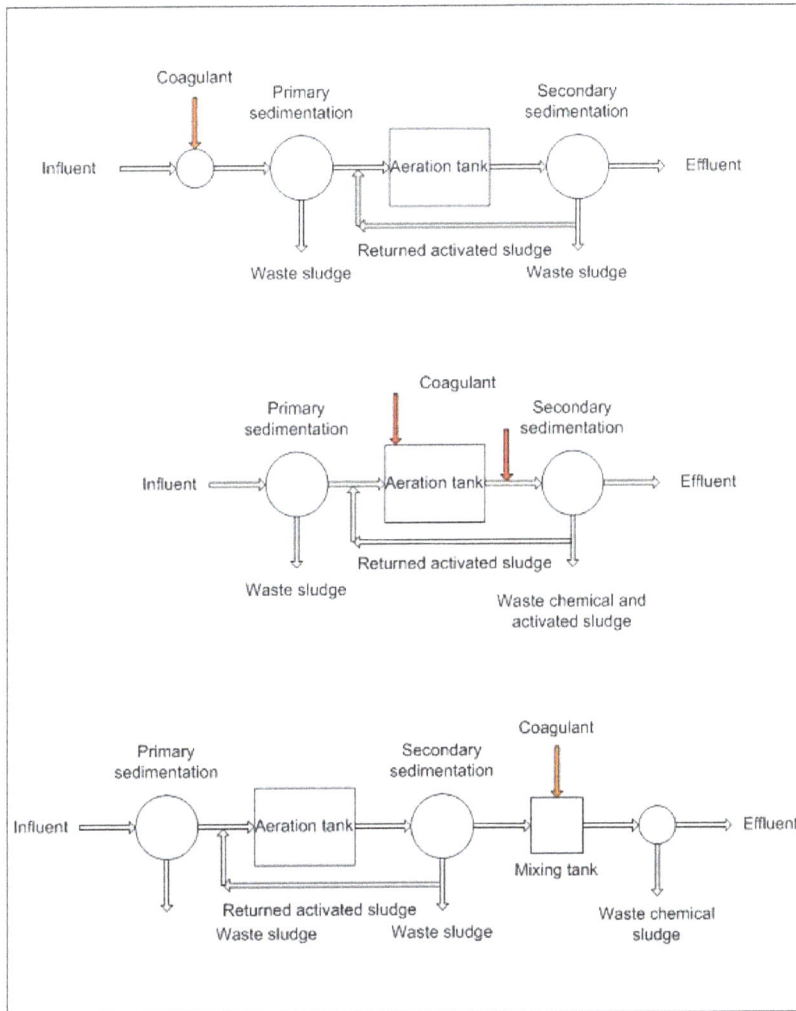

The coprecipitation process is particularly suitable for active sludge plants, where the chemicals are fed directly in the aeration tank or before it. The continuous sludge recirculation, together with the coagulation-flocculation and adsorption process due to active sludge, allows a reduction in chemical consumption. Moreover the costs for the plant are lower, since there is no need for big postprecipitation ponds. In this process the chemical added are only iron and aluminium, lime is added only for pH correction. Lower costs and more simplicity are contrasted by a phosphorous removal efficiency lower than with postprecipitation (below 85%). The phosphorous concentration in the final effluent is about 1 mg/l. Another disadvantage is that biological and chemical sludge are mixed, so they cannot be used separately in next stages. Mixed sludges need bigger sedimentation tanks than activated sludge.

Biological Processes

Over the past 20 years, several biological suspended growth process configurations have been used to accomplish biological phosphorous removal. The most important are shown in the following picture.

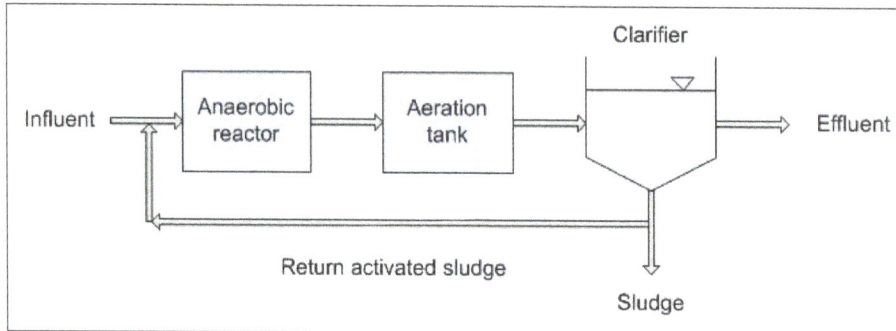

The principal advantages of biological phosphorous removal are reduced chemical costs and less sludge production as compared to chemical precipitation.

In the biological removal of phosphorous, the phosphorous in the influent wastewater is incorporated into cell biomass, which is subsequently removed from the process as a result of sludge wasting. The reactor configuration provides the P accumulating organisms (PAO) with a competitive advantage over other bacteria. So PAO are encouraged to grow and consume phosphorous. The reactor configuration in comprised of an anaerobic tank and an activated sludge activated tank. The retention time in the anaerobic tank is about 0.50 to 1.00 hours and its contents are mixed to provide contact with the return activated sludge and influent wastewater.

In the anaerobic zone: Under anaerobic conditions, PAO assimilate fermentation products (i.e. volatile fatty acids) into storage products within the cells with the concomitant release of phosphorous from stored polyphosphates. Acetate is produced by fermentation of bsCOD, which is dissolved degradable organic material that can be easily assimilated by the biomass. Using energy available from stored polyphosphates, the PAO assimilate acetate and produce intracellular polyhydroxybutyrate (PHB) storage products. Concurrent with the acetate uptake is the release of orthophosphates, as well as magnesium, potassium, calcium cations. The PHB content in the PAO increases as the polyphosphate decreases.

In the aerobic zone: energy is produced by the oxidation of storage products and polyphosphate storage within the cell increases. Stored PHB is metabolized, providing energy from oxidation and carbon for new cell growth. Some glycogen is produced from PHB metabolism. The energy released from PHB oxidation is used to form polyphosphate bonds in cell storage. The soluble orthophosphate is removed from solution and incorporated into polyphosphates within the bacterial cell. PHB utilisation also enhances cell growth and this new biomass with high polyphosphate storage accounts for phosphorous removal. As a portion of the biomass is wasted, the stored phosphorous is removed from the biotreatment reactor for ultimate disposal with the waste sludge.

The amount of phosphorous removed by biological storage can be estimated from the amount of bsCOD that is available in the wastewater influent. Better performance for BPR systems is achieved when bsCOD acetate is available at a steady rate.

Ultraviolet Germicidal Irradiation

Ultraviolet germicidal irradiation (UVGI) is a sterilization method that uses ultraviolet (UV) light at sufficiently short wavelength to break down micro-organisms. It is used in a variety of applications, such as food, air and water purification. UV has been a known mutagen at the cellular level for more than 100 years. The 1903 Nobel Prize for Medicine was awarded to Niels Finsen for his use of UV against tuberculosis.

UVGI utilises the short wavelength of UV that is harmful to forms of life at the micro-organic level. It is effective in destroying the nucleic acids in these organisms so that their DNA is disrupted by the UV radiation. This removes their reproductive capabilities and kills them.

The wavelength of UV that causes this effect is rare on Earth as its atmosphere blocks it. Using a UVGI device in certain environments like circulating air or water systems creates a deadly effect on micro-organisms such as pathogens, viruses and moulds that are in these environments. Coupled with a filtration system, UVGI can remove harmful micro-organisms from these environments.

The application of UVGI to sterilization has been an accepted practice since the mid-20th century. It has been used primarily in medical sanitation and sterile work facilities. Increasingly it was employed to sterilize drinking and wastewater, as the holding facilities were enclosed and could be circulated to ensure a higher exposure to the UV. In recent years UVGI has found renewed application in air sanitisation.

Method of Operation

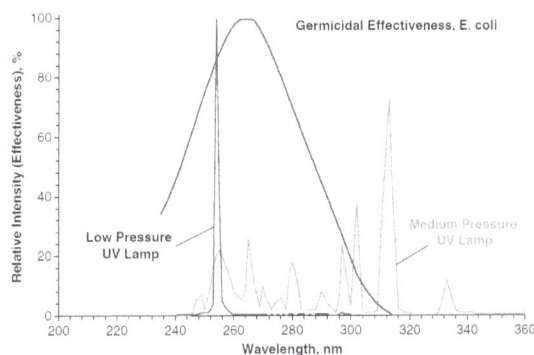

Low-pressure & medium-pressure mercury lamps compared to
e.coli germicidal effectiveness curve. Ultraviolet Germicidal
Irradiation Handbook

UV light is electromagnetic radiation with wavelengths shorter than visible light. UV can be separated into various ranges, with short-wavelength UV (UVC) considered "germicidal UV". At certain wavelengths, UV is mutagenic to bacteria, viruses and other microorganisms. Particularly at wavelengths around 260 nm–270 nm, UV breaks molecular bonds within microorganismal DNA, producing thymine dimers that can kill or disable the organisms.

- Mercury-based lamps emit UV light at the 253.7 nm line.

- Ultraviolet Light Emitting Diodes (UV-C LED) lamps emit UV light at selectable wavelengths between 255 and 280 nm.

- Pulsed-xenon lamps emit UV light across the entire UV spectrum with a peak emission near 230nm.

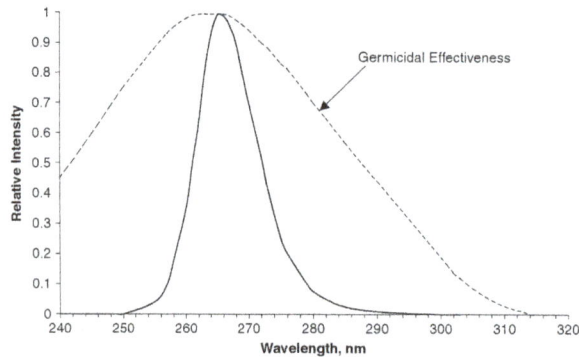

UV-C LED emitting 265 nm compared to e.coli germicidal effectiveness curve. Ultraviolet Germicidal Irradiation Handbook

This process is similar to the effect of longer wavelengths (UVB) producing sunburn in humans. Microorganisms have less protection against UV, and cannot survive prolonged exposure to it.

A UVGI system is designed to expose environments such as water tanks, sealed rooms and forced air systems to germicidal UV. Exposure comes from germicidal lamps that emit germicidal UV at the correct wavelength, thus irradiating the environment. The forced flow of air or water through this environment ensures exposure.

Effectiveness

The effectiveness of germicidal UV depends on the length of time a microorganism is exposed to UV, the intensity and wavelength of the UV radiation, the presence of particles that can protect the microorganisms from UV, and a microorganism's ability to withstand UV during its exposure.

In many systems, redundancy in exposing microorganisms to UV is achieved by circulating the air or water repeatedly. This ensures multiple passes so that the UV is effective against the highest number of microorganisms and will irradiate resistant microorganisms more than once to break them down.

"Sterilization" is often misquoted as being achievable. While it is theoretically possible in a controlled environment, it is very difficult to prove and the term "disinfection" is generally used by companies offering this service as to avoid legal reprimand. Specialist companies will often advertise a certain log reduction e.g., 99.9999% effective, instead of sterilization. This takes into consideration a phenomenon known as light and dark repair (photoreactivation and base excision repair, respectively), in which a cell can repair DNA that has been damaged by UV light.

The effectiveness of this form of disinfection depends on line-of-sight exposure of the microorganisms to the UV light. Environments where design creates obstacles that block the UV light are not as effective. In such an environment, the effectiveness is then reliant on the placement of the UVGI system so that line of sight is optimum for disinfection.

Dust and films coating the bulb lower UV output. Therefore, bulbs require periodic cleaning and replacement to ensure effectiveness. The lifetime of germicidal UV bulbs varies depending on design. Also, the material that the bulb is made of can absorb some of the germicidal rays.

Lamp cooling under airflow can also lower UV output; thus, care should be taken to shield lamps from direct airflow, or to add additional lamps to compensate for the cooling effect.

Increases in effectiveness and UV intensity can be achieved by using reflection. Aluminum has the highest reflectivity rate versus other metals and is recommended when using UV.

One method for gauging UV effectiveness in water disinfection applications is to compute UV dose. The U.S. EPA publishes UV dosage guidelines for water treatment applications. UV dose cannot be measured directly but can be inferred based on the known or estimated inputs to the process:

- Flow rate (contact time)

- Transmittance (light reaching the target)

- Turbidity (cloudiness)

- Lamp age or fouling or outages (reduction in UV intensity)

In air and surface disinfection applications the UV effectiveness is estimated by calculating the UV dose which will be delivered to the microbial population. The UV dose is calculated as follows:

UV dose $\mu Ws/cm^2$ = UV intensity $\mu W/cm^2$ x Exposure time (seconds)

The UV intensity is specified for each lamp at a distance of 1 meter. UV intensity is inversely proportional to the square of the distance so it decreases at longer distances. Alternatively, it rapidly increases at distances shorter than 1m. In the above formula the UV intensity must always be adjusted for distance unless the UV dose is calculated at exactly 1m from the lamp. Also, to ensure effectiveness the UV dose must be calculated at the end of lamp life (EOL is specified in number of hours when the lamp is expected to reach 80% of its initial UV output) and at the furthest distance from the lamp on the periphery of the target area. Some *shatter-proof* lamps are coated with a fluorated ethylene polymer to contain glass shards and mercury in case of breakage; this coating reduces UV output by as much as 20%.

To accurately predict what UV dose will be delivered to the target the UV intensity, adjusted for distance, coating and end of lamp life, will be multiplied by the exposure time. In static applications the exposure time can be as long as needed for an effective UV dose to be reached. In case of rapidly moving air, in AC air ducts for example, the exposure time is short so the UV intensity must be increased by introducing multiple UV lamps or even banks of lamps. Also, the UV installation must be located in a long straight duct section with the lamps perpendicular to the air flow to maximize the exposure time.

These calculations actually predict the UV fluence and it is assumed that the UV fluence will be equal to the UV dose. The UV dose is the amount of germicidal UV energy absorbed by a microbial population over a period of time. If the microorganisms are planktonic (free floating) the UV fluence will be equal the UV dose. However, if the microorganisms are protected by mechanical particles, such as dust and dirt, or have formed biofilm a much higher UV fluence will be needed for an effective UV dose to be introduced to the microbial population.

Inactivation of Microorganisms

The degree of inactivation by ultraviolet radiation is directly related to the UV dose applied to the water. The dosage, a product of UV light intensity and exposure time, is usually measured in microjoules per square centimeter, or equivalently as microwatt seconds per square centimeter ($\mu W \cdot s/cm^2$). Dosages for a 90% kill of most bacteria and viruses range from 2,000 to 8,000 $\mu W \cdot s/cm^2$. Larger parasites such as cryptosporidium require a lower dose for inactivation. As a result, the U.S. Environmental Protection Agency has accepted UV disinfection as a method for drinking water plants to obtain cryptosporidium, giardia or virus inactivation credits. For example, for one-decimal-logarithm reduction of cryptosporidium, a minimum dose of 2,500 $\mu W \cdot s/cm^2$ is required based on the U.S. EPA UV Guidance Manual published in 2006.

Creating UVGI

Germicidal UV is delivered by a mercury-vapor lamp that emits UV at the germicidal wavelength. Mercury vapour emits at 254nm. Many germicidal UV bulbs use special transformers to ensure even electrical flow to the bulbs so the correct wavelength is maintained. Since germicidal UV has a narrow bandwidth, power fluctuations will render intended irradiating environments ineffective. In some cases, UVGI electrodeless lamps can be energised with microwaves, giving very long stable life and other advantages. This is known as 'Microwave UV.'

There are several different types of germicidal lamps: - Low-pressure UV lamps offer high efficiencies (approx 35% UVC) but lower power, typically 1 W/cm³ power density. - Amalgam UV lamps are a high-power version of low-pressure lamps. They operate at higher temperatures and have a lifetime of up to 16,000 hours. Their efficiency is slightly lower than that of traditional low-pressure lamps (approx 33% UVC output) and power density is approx 2-3 W/cm³. - Medium-pressure UV lamps have a broad and pronounced peak-line spectrum and a high radiation output but lower UVC efficiency of 10% or less. Typical power density is 30 W/cm³ or greater.

Depending on the quartz glass used for the lamp body, low-pressure and amalgam UV lamps emit light at 254 nm and 185 nm (for oxidation).

185 nm light is used to generate ozone.

Strengths and Weaknesses

Advantages

UV water treatment devices can be used for well water and surface water disinfection. UV treatment compares favorably with other water disinfection systems in terms of cost, labor, and the need for technically trained personnel for operation. Water chlorination treats larger organisms and offers residual disinfection, but these systems are expensive because they need special operator training and a steady supply of a potentially hazardous material. Finally, boiling of water is the most reliable treatment method but it demands labor, and imposes a high economic cost. UV treatment is rapid and, in terms of primary energy use, approximately 20,000 times more efficient than boiling.

Disadvantages

UV disinfection is most effective for treating high-clarity, purified reverse osmosis distilled water. Suspended particles are a problem because microorganisms buried within particles are shielded from the UV light and pass through the unit unaffected. However, UV systems can be coupled with a pre-filter to remove those larger organisms that would otherwise pass through the UV system unaffected. The pre-filter also clarifies the water to improve light transmittance and therefore UV dose throughout the entire water column. Another key factor of UV water treatment is the flow rate—if the flow is too high, water will pass through without sufficient UV exposure. If the flow is too low, heat may build up and damage the UV lamp.

A disadvantage of UVGI is that while water treated by chlorination is resistant to reinfection (until the chlorine off-gasses), UVGI water is not resistant to reinfection. UVGI water must be transported or delivered in such a way as to avoid reinfection.

Technology

Lamps

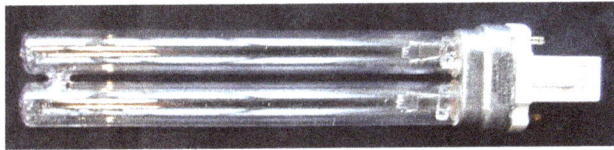

A 9 W germicidal lamp in a compact fluorescent lamp form factor

Germicidal UV for disinfection is most typically generated by a mercury-vapor lamp. Low-pressure mercury vapor has a strong emission line at 254 nm, which is within the range of wavelengths that demonstrate strong disinfection effect. The optimal wavelengths for disinfection are close to 270 nm.

Lamps are either amalgam or medium-pressure lamps. Low-pressure UV lamps offer high efficiencies (approx 35% UVC) but lower power, typically 1 W/cm power density (power per unit of arc length). Amalgam UV lamps are a higher-power version of low-pressure lamps. They operate at higher temperatures and have a lifetime of up to 16,000 hours. Their efficiency is slightly lower than that of traditional low-pressure lamps (approx 33% UVC output) and power density is approximately 2–3 W/cm. Medium-pressure UV lamps have a broad and pronounced peak-line spectrum and a high radiation output but lower UVC efficiency of 10% or less. Typical power density is 30 W/cm^3 or greater.

Depending on the quartz glass used for the lamp body, low-pressure and amalgam UV emit radiation at 254 nm and also at 185 nm, which has chemical effects. UV radiation at 185 nm is used to generate ozone.

The UV lamps for water treatment consist of specialized low-pressure mercury-vapor lamps that produce ultraviolet radiation at 254 nm, or medium-pressure UV lamps that produce a polychromatic output from 200 nm to visible and infrared energy. The UV lamp never contacts the water; it is either housed in a quartz glass sleeve inside the water chamber or mounted external to the water which flows through the transparent UV tube. Water passing through the flow chamber is exposed to UV rays which are absorbed by suspended solids, such as microorganisms and dirt, in the stream.

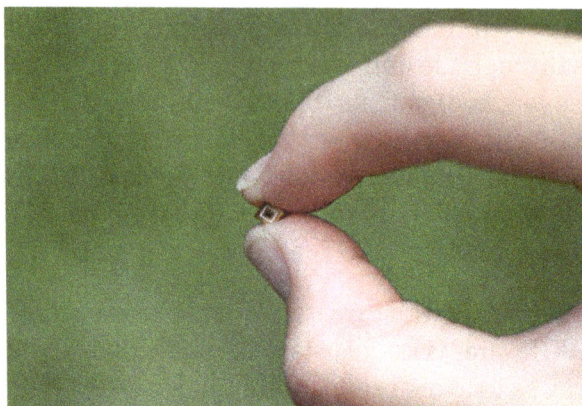

Compact and versatile options with UV-C LEDs

Light Emitting Diodes (LEDs)

Recent developments in LED technology have led to commercially available UV-C LEDs. UV-C LEDs use semiconductors to emit light between 255 nm-280 nm. The wavelength emission is tuneable by adjusting the material of the semiconductor. The reduced size of LEDs opens up options for small reactor systems allowing for point-of-use applications and integration into medical devices. Low power consumption of semiconductors introduce UV disinfection systems that utilized small solar cells in remote or Third World applications.

Water Treatment Systems

Sizing of a UV system is affected by three variables: flow rate, lamp power, and UV transmittance in the water. Manufacturers typically developed sophisticated Computational Fluid Dynamics (CFD) models validated with bioassay testing. This involves testing the UV reactor's disinfection performance with either MS2 or T1 bacteriophages at various flow rates, UV transmittance, and power levels in order to develop a regression model for system sizing. For example, this is a requirement for all drinking water systems in the United States per the EPA UV Guidance Manual.[5-2]

The flow profile is produced from the chamber geometry, flow rate, and particular turbulence model selected. The radiation profile is developed from inputs such as water quality, lamp type (power, germicidal efficiency, spectral output, arc length), and the transmittance and dimension of the quartz sleeve. Proprietary CFD software simulates both the flow and radiation profiles. Once the 3D model of the chamber is built, it is populated with a grid or mesh that comprises thousands of small cubes.

Points of interest—such as at a bend, on the quartz sleeve surface, or around the wiper mechanism—use a higher resolution mesh, whilst other areas within the reactor use a coarse mesh. Once the mesh is produced, hundreds of thousands of virtual particles are "fired" through the chamber. Each particle has several variables of interest associated with it, and the particles are "harvested" after the reactor. Discrete phase modeling produces delivered dose, head loss, and other chamber-specific parameters.

When the modeling phase is complete, selected systems are validated using a professional third party to provide oversight and to determine how closely the model is able to predict the reality of system performance. System validation uses non-pathogenic surrogates such as MS 2 phage or *Bacillus subtilis* to determine the Reduction Equivalent Dose (RED) ability of the reactors. Most systems are validated to deliver 40 mJ/cm² within an envelope of flow and transmittance.

To validate effectiveness in drinking-water systems, the method described in the EPA UV Guidance Manual is typically used by the U.S., whilst Europe has adopted Germany's DVGW 294 standard. For wastewater systems, the NWRI/AwwaRF Ultraviolet Disinfection Guidelines for Drinking Water and Water Reuse protocols are typically used, especially in wastewater reuse applications.

Potential Dangers

At certain wavelengths (including UVC) UV is harmful to humans and other forms of life. In most UVGI systems the lamps are shielded or are in environments that limit exposure, such as a closed water tank or closed air circulation system, often with interlocks that automatically shut off the UV lamps if the system is opened for access by human beings. Limited exposure mitigates the risk of danger.

In human beings, skin exposure to germicidal wavelengths of UV light can produce sunburn and (in some cases) skin cancer. Exposure of the eyes to this UV radiation can produce extremely painful inflammation of the cornea and temporary or permanent vision impairment, up to and including blindness in some cases. UV can damage the retina of the eye.

Another potential danger is the UV production of ozone. UVC light from the sun is partly responsible for the earth's ozone layer in the stratosphere, but ozone at the atmospheric level can be harmful to a person's health. The United States Environmental Protection Agency designated .05 parts per million (ppm) of ozone to be a safe level.

UV-C radiation is able to break-down chemical bonds. This leads to rapid ageing of plastics (insulations, gasket) and other materials. Note that plastics sold to be "UV-resistant" are tested only for UV-B, as UV-C doesn't normally reach the surface of the Earth.

Uses for UVGI

Air Purification

UVGI can be used to sterilize air that passes UV lamps via forced air. Air purification UVGI systems can be freestanding units with shielded UV lamps that use a fan to force air past the UV light. Other systems are installed in forced air systems so that the circulation for the premises moves micro-organisms past the lamps. Key to this form of sterilization is placement of the UV lamps and a good filtration system to remove the dead micro-organisms. For example, forced air systems by design impede line of sight, thus creating areas of the environment that will be shaded from the UV light. However, a UV lamp placed at the coils and drain pan of cooling system will keep micro-organisms from forming in these naturally damp places. The most effective method for treating the air rather than the coils is in-line duct systems, these systems are placed in the center of the duct and parallel to the air flow.

Water Purification

Water purification via UVGI is used in most water sterilization processes, such as purification, detoxification and disinfection. Its use in wastewater treatment is replacing chlorination due to that chemical's toxic by-products. A disadvantage is that water treated by chlorination is resistant to reinfection, where UVGI water must be transported and delivered in such a way as to avoid contamination. Individual waste streams to be treated by UVGI must be tested to ensure that the method will be effective due to potential interferences such as suspended solids, dyes or other substances that may block or absorb the UV radiation.

"UV units to treat small batches (1 to several liters) or low flows (1 to several liters per minute) of water at the community level are estimated to have costs of 0.02 US$ per 1000 liters of water, including the cost of electricity and consumables and the annualized capital cost of the unit." (WHO).

Laboratory Hygiene

UVGI is often used to sterilize equipment such as safety goggles, instruments, pipettors, and other devices. Lab personnel also sterilize glassware and plasticware this way. Microbiology laboratories use UVGI to sterilize surfaces inside biological safety cabinets ("hoods") between uses.

Septic Tank

A septic tank collects and treats wastewater at a property that is not connected to the mains sewer system.

Installed underground, a septic tank makes use of natural processes to treat the sewage it stores. Usually made up of two chambers or compartments, the tank receives wastewater from an inlet pipe.

Wastewater enters the first chamber and separates over time, with solids settling at the bottom, oils and greases forming a layer of scum at the top, and a layer of relatively clear water remaining in the middle.

This clarified wastewater enters the second chamber. It then exits via an outlet pipe into a septic drain field, also known as a seepage field or leach field. The remaining scum and solids in the tank are broken down by naturally occurring bacteria and what is left should be professionally removed periodically.

The septic tank was patented in London around 1900. Websters Dictionary defines the septic tank as "A tank in which waste matter is decomposed through bacterial action." The modern septic tank is a watertight box usually made of precast concrete, concrete blocks, or reinforced fiberglass.

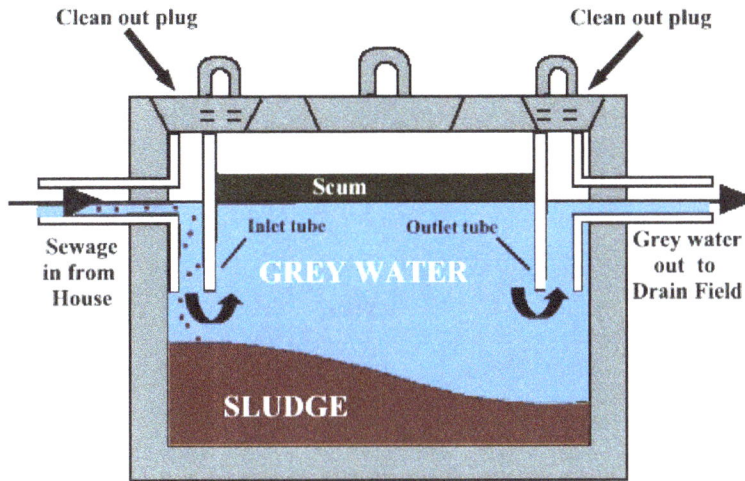

The septic system is a small, on-site treatment and disposal system buried in the ground. the septic system has two essential parts: (1) the septic tank and (2) the soil absorption area. When household waste enters the septic tank several things occur:

1. Organic solid material floats to the surface and forms a layer of what is commonly called "scum." Bacteria in the septic tank biologically convert this material to liquid.

2. Inorganic or inert solid materials and the by-products of bacterial digestions sink to the bottom of the tank and form a layer commonly known as "sludge."

3. Only clear water should exist between the scum and sludge layers. It is this clear water – And only this clear water – that should overflow into the soil absorption area.

Septic Tank with Baffle

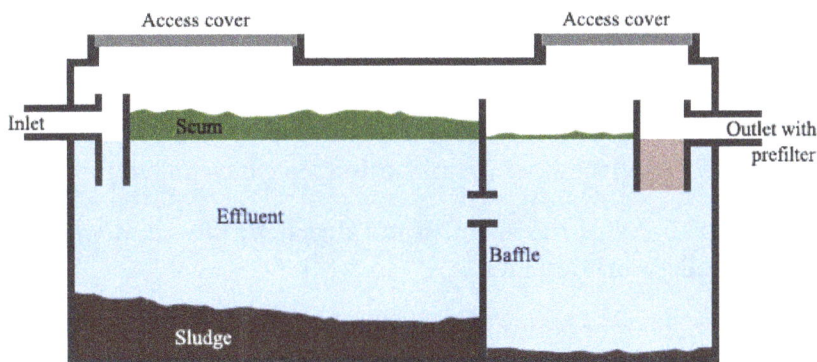

Fosse Septique

Solid material overflowing into the soil absorption area should be avoided at all costs. It is this solids overflow that clogs soil pores and causes system to fail. two main factors cause solid material to build up enough to overflow: (1) bacterial deficency, and (2) lack of sludge removal.

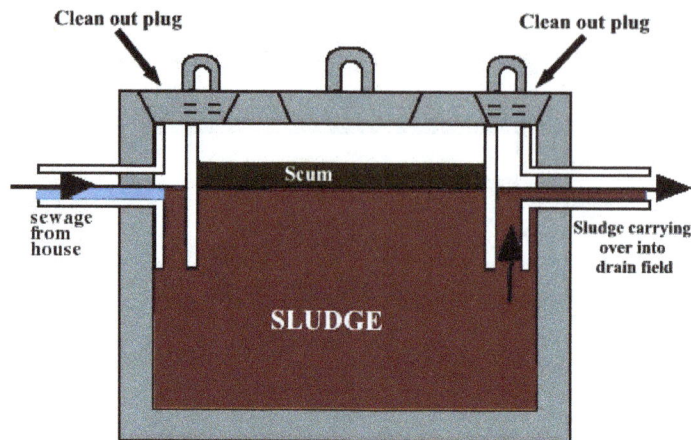

Bacteria must be present in the septic tank to break down and digest the organic solids. Normal household waste probides enough bacteria to digest the solids unless any harm is done to the bacteria. Bacteria are very sensitive to environmental changes. Check teh lables of products you normally use in home. Products carrying harsh warnings such as "Harmful or Fatal if Swallowed" will harm bacteria.

- Detergents

- Bleaches

- Cleaning compounds

- Disinfectants

- Acids

- Toilet cleaners

- Polishes

- Caustic drain openers

People rarely think of the effect of these products on the septic tank system when the products go down the drain. What kind of effect to you think anti-septics have on your septic tank?

Bacteria must be present to digest the scum. If not digested, the scum will accumulate untill it overflows, clogging the soil absorption area.

The sludge in the septic tank – inorganic and inert material – is not biodegradable and will not decompose. If not removed, the sludge will accumulate until it eventually overflows, again clogging the soil absorption area.

Emptying

Waste that is not decomposed by the anaerobic digestion must eventually be removed from the septic tank. Otherwise the septic tank fills up and wastewater containing undecomposed material discharges directly to the drainage field. Not only is this detrimental for the environment but, if the sludge overflows the septic tank into the leach field, it may clog the leach field piping or decrease the soil porosity itself, requiring expensive repairs.

A vacuum truck used to empty septic tanks in Germany

When a septic tank is emptied, the accumulated sludge (septage, also known as fecal sludge) is pumped out of the tank by a vacuum truck. How often the septic tank must be emptied depends on the volume of the tank relative to the input of solids, the amount of indigestible solids, and the ambient temperature (because anaerobic digestion occurs more efficiently at higher temperatures), as well as usage, system characteristics and the requirements of the relevant authority. Some health authorities require tanks to be emptied at prescribed intervals, while others leave it up to the decision of an inspector. Some systems require pumping every few years or sooner, while others may be able to go 10–20 years between pumpings. An older system with an undersize tank that is being used by a large family will require much more frequent pumping than a new system used by only a few people. Anaerobic decomposition is rapidly restarted when the tank is refilled.

Services for de-sludging tend to empty a septic tank completely, i.e. take out all septage, while the actual requirement is removal of settled solids, and even this purposefully incompletely so as to leave at least some of the microbial populations in place to continue the anaerobic degradation processes that take place in a septic tank.

Maintenance

A septic tank before installation, with manhole cover on top

The same tank partially installed in the ground

Like any system, a septic system requires maintenance. The maintenance of a septic system is often the responsibility of the resident or property owner. Some forms of abuse or neglect include the following:

User's Actions

- Excessive disposal of cooking oils and grease can cause the inlet drains to block. Oils and grease are often difficult to degrade and can cause odor problems and difficulties with the periodic emptying.

- Flushing non-biodegradable waste items down the toilet such as cigarette butts, cotton buds/swabs or menstrual hygiene products and condoms can cause a septic tank to clog and fill rapidly, so these materials should not be disposed of in that manner. The same applies when the toilet is connected to a sewer rather than a septic tank.

- Using the toilet for disposal of food waste can cause a rapid overload of the system with solids and contribute to failure.

- Certain chemicals may damage the components of a septic tank or kill the bacteria needed in the septic tank for the system to operate properly, such as pesticides, herbicides, materials with high concentrations of bleach or caustic soda (lye), or any other inorganic materials such as paints or solvents.

- Using water softeners - the brine discharge from water softeners may harm the bacteria responsible for breaking down the wastewater. Usually the brine is however sufficiently diluted with other wastewater that it does not adversely affect the septic system.

Other Factors

- Roots from trees and shrubbery protruding above the tank or drainfield may clog and/or rupture them. Trees that are directly within the vicinity of a concrete septic tank have the potential to penetrate the tank as the system ages and the concrete begins to develop cracks and small leaks. Tree roots can cause serious flow problems due to plugging and blockage of drain pipes, added to which the trees themselves tend to expand extremely vigorously due to the ready supply of nutrients from the septic system.

- Playgrounds and storage buildings may cause damage to a tank and the drainage field. In addition, covering the drainage field with an impermeable surface, such as a driveway or parking area, will seriously affect its efficiency and possibly damage the tank and absorption system.

- Excessive water entering the system may overload it and cause it to fail.

- Very high rainfall, rapid snowmelt, and flooding from rivers or the sea can all prevent a drain field from operating, and can cause flow to back up, interfering with the normal operation of the tank. High winter water tables can also result in groundwater flowing back into the septic tank.

- Over time, biofilms develop on the pipes of the drainage field, which can lead to blockage. Such a failure can be referred to as "biomat failure".

Septic Tank Additives

Septic tank additives have been promoted by some manufacturers with the aim to improve the effluent quality from septic tanks, reduce sludge build-up and to reduce odors. However, these additives—which are commonly based on "effective microorganisms"—are usually costly in the longer term and fail to live up to expectations. It has been estimated that in the U.S. more than 1,200 septic system additives were available on the market in 2011. However, very little peer-reviewed and replicated field research exists regarding the efficacy of these biological septic tank additives.

Environmental Concerns

While a properly maintained and located septic tank does not pose any more environmental problems than centralized municipal sewage treatment, certain problems can arise with septic tanks in unsuitable locations. Since septic systems require large drainfields, they are not suitable for densely built areas.

Odor and Gas Emissions

Some constituents of wastewater, especially sulfates, under the anaerobic conditions of septic tanks, are reduced to hydrogen sulfide, a pungent and toxic gas. Methane may also be released. Nitrates and organic nitrogen compounds can be reduced to ammonia. Because of the anaerobic conditions, fermentation processes take place, which may generate carbon dioxide and/or methane.

Nutrients in the Effluent

Septic tanks by themselves are ineffective at removing nitrogen compounds that have potential to cause algal blooms in waterways into which affected water from a septic system finds its way. This can be remedied by using a nitrogen-reducing technology, or by simply ensuring that the leach field is properly sited to prevent direct entry of effluent into bodies of water.

The fermentation processes cause the contents of a septic tank to be anaerobic with a low redox potential, which keeps phosphates in a soluble and, thus, mobilized form. Phosphates discharged from a septic tank into the environment can trigger prolific plant growth including algal blooms, which can also include blooms of potentially toxic cyanobacteria.

The soil's capacity to retain phosphorus is usually large enough to handle the load through a normal residential septic tank. An exception occurs when septic drain fields are located in sandy or coarser soils on property adjoining a water body. Because of limited particle surface area, these soils can become saturated with phosphates. Phosphates will progress beyond the treatment area, posing a threat of eutrophication to surface waters.

Groundwater Pollution

In areas with high population density, groundwater pollution beyond acceptable limits may occur. Some small towns are experiencing the costs of building very expensive centralized wastewater treatment systems because of this problem, owing to the high cost of extended collection systems. To reduce residential development which might increase the demand to construct an expensive centralized sewerage system, building moratoriums and limits on the subdivision of property are often imposed. Ensuring existing septic tanks are functioning properly can also be helpful for a limited time, but becomes less effective as a primary remediation strategy as population density increases.

Surface Water Pollution

In areas adjacent to water bodies with fish or shellfish intended for human consumption, improperly maintained and failing septic systems contribute to pollution levels that can force harvest restrictions and/or commercial or recreational harvest closures.

Septic Tank Pumping

A large majority of on-lot sewage systems have septic tanks. The question of how often a septic tank should be pumped has been debated for many years. On the one hand you will find homeowners who claim they have never pumped their septic tank and it "seems" to work perfectly. On the other hand, in an attempt to create a uniform pumping policy, regulators have come down on the conservative side and have stated that all septic tanks should be pumped every two or three years.

How Frequent should a Septic Tank be Pumped?

The frequency of pumping depends on several factors:

- Capacity of septic tank.

- Daily volume of wastewater added to the septic tank.

- Amount of solids in wastewater stream. It should be noted that there are several classes of solids that are commonly put into a septic tank. These include (1) biodegradable "organic" solids such as feces, (2) slowly biodegradable "organic" solids such as toilet paper and cellulosic compounds, which take a long time to biodegrade in the septic tank, and (3) non-biodegradable solids such as kitty litter, plastics, etc., which do not biodegrade and quickly fill the septic tank. Reducing the amount of slowly biodegradable organics and non-biodegradable waste added to your septic tank will greatly reduce the rate at which solids accumulate in the tank.

Another contributor to how quickly a septic tank will fill with solids is life style. The two most important life-style issues related to septic tank performance are:

1. Water usage in the home, and

2. Age of the residents.

Homes with growing families including children ranging from small children to teenagers usually use more water and put more solids into the septic tank. On the other hand, empty nesters, and especially the elderly tend to use much less water and put smaller amounts of solids into septic tanks.

Another important consideration regarding how often a septic tank should be pumped is timing. As stated earlier, as a septic tank fills with solids, these solids tend to be carried from the tank to the soil absorption area, especially from tanks that do not have exit filters. As more solids accumulate in the absorption area, these solids begin to clog the soil and restrict the movement of wastewater into the soil. By the time sewage has backed up into the home, the soil absorption area is clogged with a nearly impermeable biomat and flooded with wastewater because the soil is no longer able to absorb the wastewater produced on a daily basis. Removal of these biomats is usually expensive and time consuming. Pumping the septic tank will not remove the biomat. Removing the biomat requires that you pump the wastewater ponded in the soil absorption area. Then the pump-access hole to the absorption area should be left open for several days. Once the absorption area is free of water and has become aerated, the biomat usually decomposes in a few days. When the absorption area is pumped, the septic tank should also be pumped, thus enhancing the development of the aerated conditions in the absorption area.

Time to Pump your Septic Tank

So, how does one decide how often a septic tank should be pumped? We know homes that put large amounts of non-biodegradable and slowly biodegradable organics into the septic tank need to pump more often. We also know that the septic tank should be pumped before the captured solids accumulate to the point where these solids begin being carried with the tank effluent to the absorption area. There are two relatively safe approaches to deciding when (or how often) to pump your septic tank. One is to just have it pumped every two or three years. The other is to open the access port to the first chamber once every year and insert a long pole to the bottom of the tank and withdraw it. You can see the depth of sludge by the darkness on the pole. If the sludge is more than a third of the tank depth, it is time to have it pumped. Most homeowners are better off just having their tank pumped every two or three years.

The Pumping Process

Septic tank pumping and haul contractors can pump your septic tank. It is a good idea to be on hand to ensure that it is done properly. To extract all the material from the tank, the scum layer must be broken up and the sludge layer mixed with the liquid portion of the tank. This is usually done by alternately pumping liquid from the tank and re-injecting it into the bottom of the tank. The septic tank must be pumped through the two large central access ports (manholes), not the small inlet or outlet inspection ports located above each baffle. Pumping a tank through the baffle inspection ports can damage the baffles and yield incomplete removal of sludge and scum.

The use of additives in septic tanks to reduce the sludge volume or as a substitute for pumping is not recommended. In fact, relying on additives rather than conventional tank pumping may result in failure of the entire septic system.

When you have your septic tank pumped, an additional step may help keep your septic system functioning properly for a long time. Most companies that pump septic tanks also have a certified PSMA Inspector in their company. This inspector can tell you if your septic tank needs repair or if other components of your septic system need maintenance.

To facilitate future cleaning and inspection, install risers from the central access ports and inspection ports to the soil surface. Also mark the location of the tank, so it can be easily located for future pumping.

Schedule Septic Tank Pumping

Homeowners should get in the habit of having the septic tank pumped. If you are able and willing to have your septic tank pumped on a routine basis (such as every two or three years), it may be possible to further enhance the effectiveness of your entire on-lot wastewater disposal system. Research at Penn State has shown that your soil absorption system will benefit from periodic resting (a period during which no wastewater is added to the absorption area). To get the greatest benefit from pumping your septic tank, it is recommended that you have your septic tank pumped every two to three years on the day before you, and your family, leave for your summer vacation. This means the whole system, especially the soil absorption area, will have the opportunity to dry out and any partially decomposed organic waste (biomat) that may have developed in the soil absorption area will quickly decompose in the absence of water.

Aerobic Treatment Unit

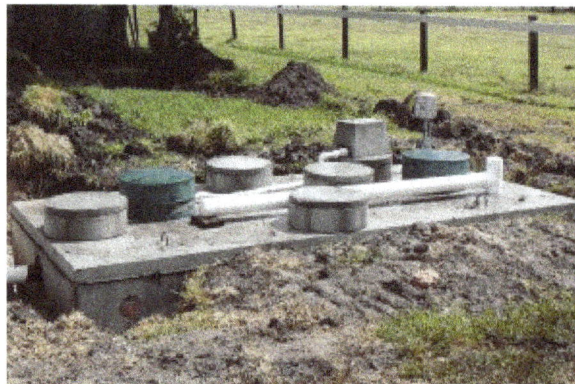

An aerobic treatment unit (ATU) consists of several processes that function together to provide a high quality effluent. These are gross solids (trash) removal, aeration, clarification, and sludge return. These processes are generally contained within separate chambers of a single tank. A series of tanks can be configured to have wastewater pass through an aerobic treatment train.

ATUs use biological processes to transform both dissolved and solid constituents into gases, cell mass, and non-degradable material. An important feature of the biological process is the synthesis and separation of microbial cells from the treated effluent.

The treatment process involves a variety of aerobic and facultative microorganisms living together that can decompose a broad range of materials. The organisms live in an aerobic environment where free oxygen is available for their respiration. ATUs can be used to remove substantial amounts of BOD5 and TSS that are not removed by simple sedimentation in a conventional septic system.

The biological process also involves the nitrification of ammonia in the wastewater and the reduction of pathogenic organisms. Nitrification is the breakdown of ammonia (NH_3+) to nitrate (NO_3-) by microorganisms in aerobic conditions.

ATUs, which are certified as Class I aerobic systems, treat wastewater well enough to be used in conjunction with spray systems, which distribute treated wastewater over lawns. Combined with disinfection, they are the most common way to treat wastewater for spray systems.

Process

The ATS process generally consists of the following phases:

- Pre-treatment stage to remove large solids and other undesirable substances.

- Aeration stage, where aerobic bacteria digest biological wastes.

- Settling stage allows undigested solids to settle. This forms a sludge that must be periodically removed from the system.

- Disinfecting stage, where chlorine or similar disinfectant is mixed with the water, to produce an antiseptic output. Another option is UV disinfection, where the water is exposed to UV light inside of a UV disinfection unit.

The disinfecting stage is optional, and is used where a sterile effluent is required, such as cases where the effluent is distributed above ground. The disinfectant typically used is tablets of calcium hypochlorite, which are specially made for waste treatment systems. The tablets are intended to break down quickly in sunlight. Stabilized forms of chlorine persist after the effluent is dispersed, and can kill plants in the leach field.

Since the ATS contains a living ecosystem of microbes to digest the waste products in the water, excessive amounts of items such as bleach or antibiotics can damage the ATS environment and reduce treatment effectiveness. Non-digestible items should also be avoided, as they will build up in the system and require more frequent sludge removal.

Types of Aerobic Treatment Systems

Small scale aerobic systems generally use one of two designs, fixed-film systems, or continuous flow, suspended growth aerobic systems (CFSGAS). The pre-treatment and effluent handling are similar for both types of systems, and the difference lies in the aeration stage.

Fixed Film Systems

Fixed film systems use a porous medium which provides a bed to support the biomass film that digests the waste material in the wastewater. Designs for fixed film systems vary widely, but fall into two basic categories (though some systems may combine both methods). The first is a system where the media is moved relative to the wastewater, alternately immersing the film and exposing it to air, while the second uses a stationary media, and varies the wastewater flow so the film is alternately submerged and exposed to air. In both cases, the biomass must be exposed to both wastewater and air for the aerobic digestion to occur. The film itself may be made of any suitable porous material, such as formed plastic or peat moss. Simple systems use stationary media, and rely on intermittent, gravity driven wastewater flow to provide periodic exposure to air and wastewater. A common moving media system is the rotating biological contactor (RBC), which uses disks rotating slowly on a horizontal shaft. Approximately 40 percent of the disks are submerged at any given time, and the shaft rotates at a rate of one or two revolutions per minute.

Continuous Flow, Suspended Growth Aerobic Systems

CFSGAS systems, as the name implies, are designed to handle continuous flow, and do not provide a bed for a bacterial film, relying rather on bacteria suspended in the wastewater. The suspension and aeration are typically provided by an air pump, which pumps air through the aeration chamber, providing a constant stirring of the wastewater in addition to the oxygenation. A medium to promote fixed film bacterial growth may be added to some systems designed to handle higher than normal levels of biomass in the wastewater.

Retrofit or Portable Aerobic Systems

Another increasingly common use of aerobic treatment is for the remediation of failing or failed anaerobic septic systems, by retrofitting an existing system with an aerobic feature. This class of product, known as aerobic remediation, is designed to remediate biologically failed and failing anaerobic distribution systems by significantly reducing the biochemical oxygen demand (BOD5) and total suspended solids (TSS) of the effluent. The reduction of the BOD5 and TSS reverses the developed bio-mat. Further, effluent with high dissolved oxygen and aerobic bacteria flow to the distribution component and digest the bio-mat.Doing so on single tank systems where solids do not have anywhere to settle, or there is no a clarifying area can do damage to the field lines as the solid matter is stirred up in the tank.

Composting Toilets

Composting toilets are designed to treat only toilet waste, rather than general residential waste water, and are typically used with water-free toilets rather than the flush toilets associated with the above types of aerobic treatment systems. These systems treat the waste as a moist solid, rather than in liquid suspension, and therefore separate urine from feces during treatment to maintain the correct moisture content in the system. An example of a composting toilet is the clivus multrum (Latin for 'inclined chamber'), which consists of an inclined chamber that separates urine and feces and a fan to provide positive ventilation and prevent odors from escaping through the toilet. Within the chamber, the urine and feces are independently broken down not only by aerobic bacteria, but also by fungi, arthropods, and earthworms. Treatment times are very long, with a minimum time between removals of solid waste of a year; during treatment the volume of the solid waste is decreased by 90 percent, with most being converted into water vapor and carbon dioxide. Pathogens are eliminated from the waste by the long durations in inhospitable conditions in the treatment chamber.

Comparison to Traditional Septic Systems

The aeration stage and the disinfecting stage are the primary differences from a traditional septic system; in fact, an aerobic treatment system can be used as a secondary treatment for septic tank effluent. These stages increase the initial cost of the aerobic system, and also the maintenance requirements over the passive septic system. Unlike many other biofilters, aerobic treatment systems require a constant supply of electricity to drive the air pump increasing overall system costs. The disinfectant tablets must be periodically replaced, as well as the electrical components (air compressor) and mechanical components (air diffusers). On the positive side, an aerobic system produces a higher quality effluent than a septic tank, and thus the leach field can be smaller than that of a conventional septic system, and the output can be discharged in areas too environmentally sensitive for septic system output. Some aerobic systems recycle the effluent through a sprinkler system, using it to water the lawn where regulations approve.

Effluent Quality

Since the effluent from an ATS is often discharged onto the surface of the leach field, the quality is very important. A typical ATS will, when operating correctly, produce an effluent with less than 30 mg/liter BOD5, 25 mg/L TSS, and 10,000 cfu/mL fecal coliform bacteria. This is clean enough that it cannot support a biomat or "slime" layer like a septic tank.

ATS effluent is relatively odorless; a properly operating system will produce effluent that smells musty, but not like sewage. Aerobic treatment is so effective at reducing odors, that it is the preferred method for reducing odor from manure produced by farms.

API Oil–water Separator

The API oil/water separator is a simple gravity separation device designed by using Stokes Law to define the rise velocity of oil droplets based on their density and size. The design of the separator is based on the specific gravity difference between the oil and the wastewater because that difference is usually smaller than the specific gravity difference between the suspended solids and water. Based on that design criterion, most of the suspended solids will settle to the bottom of the separator as a sediment layer, the oil will rise to top of the separator, and the wastewater will be the middle layer between the oil on top and the solids on the bottom.

API oil/water separators are designed to remove oil droplets down to 150 micron, which is not nearly enough removal necessary to stay compliant today. Most API oil/water separators are not going to do any better than 50 parts per million as opposed to most state & federal regulations that today require below 20 ppm.

Most industries have abandoned building these large gravity settling vessels because they are simply too costly to justify. Most are built of concrete and with the high cost of labor and the need to reinforce these tanks with steel, new construction costs rise quickly. Plus API units have an excessively large footprint and are far too much of a real estate burden to consider this as a viable option.

Those who have API oil/water separators acknowledge they are easy to operate and maintain on a day-to-day basis, but when it is time to clean out the tank, the cost to pump out the sludge and clean this large space is just a nightmare. Some units have a chain and scrapper mechanism. This helps move the solids to one side to be pumped out, but the automation piece for these are usually as much as the construction of the pit itself—and require significant maintenance as well.

Description of the Design and Operation

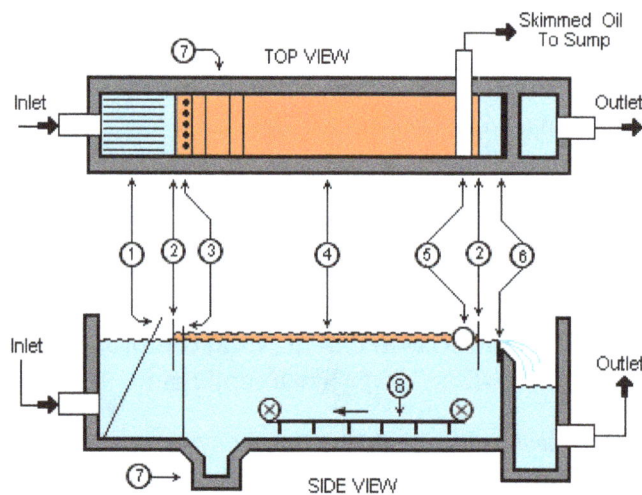

1 Trash trap (inclined rods)
2 Oil retention baffles
3 Flow distributors (vertical rods)
4 Oil layer
5 Slotted pipe skimmer
6 Adjustable overflow weir
7 Sludge sump
8 Chain and flight scraper

A typical gravimetric API separator

The API separator is a gravity separation device designed using Stokes' law principles that define the rise velocity of oil droplets based on their density, size and water properties. The design of the separator is based on the specific gravity difference between the oil and the wastewater because that difference is much smaller than the specific gravity difference between the suspended solids and water. Based on that design criterion, most of the suspended solids will settle to the bottom of the separator as a sediment layer, the oil will rise to top of the separator, and the wastewater will be the middle layer between the oil on top and the solids on the bottom. The API Design Standards, when correctly applied, make adjustments to the geometry, design and size of the separator beyond simple Stokes Law principles. This includes allowances for water flow entrance and exit turbulence losses as well as other factors. API Specification 421 requires a minimum length to width ratio of 5:1 and minimum depth-to-width ratio of 0.3:0.5.

Typically in operation of API separators the oil layer, which may contain entrained water and attached suspended solids, is continually skimmed off. This removed oily layer may be re-processing to recover valuable products, or disposed of. The heavier bottom sediment layer is removed by a chain and flight scraper (or similar device) and a sludge pump.

Design Limitations

API design separators, and similar gravity tanks, are not intended to be effective when any of the following conditions apply to the feed conditions:

- Mean Oil droplets size in the feed is less than 150 micron

- Oil density is greater than 925 kg/m3

- Suspended solids are adhering to the oil meaning the 'effective' oil density is greater than 925 kg/m3

- Water temperature less than 5 °C

- There are high levels of dissolved hydrocarbons

According to Stokes' Law, heavier oils require more retention time. In many cases where refineries have switched to heavier crude slates, the API separator's efficiency has declined.

Further Treatment of API Water Discharges

Because of performance limitations the water discharged from API type separators usually requires several further processing stages before the treated water can be discharged or reused. Further water treatment is designed to remove oil droplets smaller than 150 micron, dissolved materials and hydrocarbons, heavier oils or other contaminants not removed by the API. Secondary treatment technologies include dissolved air flotation (DAF), Anaerobic and Aerobic biological treatment, Parallel Plate Separators, Hydrocyclone, Walnut Shell Filters and Media filters.

Alternative Technologies

Plate separators, or Coalescing Plate Separators are similar to API separators, in that they are based on Stokes Law principles, but include inclined plate assemblies (also known as parallel packs). The underside of each parallel plate provides more surface for suspended oil droplets to coalesce into larger globules. Coalescing plate separators may not be effective in situation where water chemicals or suspended solids restrict or prevent oil droplets coalesce. In operation it is intended that sediment will slide down the topside of each parallel plate, however in many practical situations the sediment can adhere to the plates requiring periodic removal and cleaning. Such separators still depend upon the specific gravity between the suspended oil and the water. However, the parallel plates can enhance the degree of oil-water separation for oil droplets above 50 micron in size. Alternatively parallel plate separators are added to the design of API Separators and require less space than a conventional API separator to achieve a similar degree of separation.

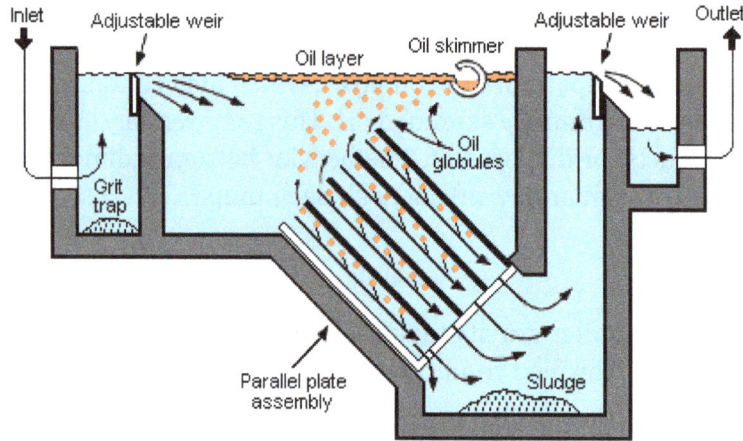

A typical parallel plate separator

Other Oil–water Separation Applications

There are other applications requiring oil-water separation. For example:

- Oily water separators (OWS) for separating oil from the bilge water accumulated in ships as required by the international MARPOL Convention.

- Oil and water separators are commonly used in electrical substations. The transformers found in substations use a large amount of oil for cooling purposes. Moats are constructed surrounding unenclosed substations to catch any leaked oil, but these will also catch rainwater. Oil and water separators therefore provide a quicker and easier cleanup of an oil leak.

Sewage Sludge Treatment

Sewage sludge consists of water and solids that can be divided into mineral and organic solids. The quantity and characteristics of sludge depend very much on the treatment processes. Most of the pollutants that enter the wastewater get adsorbed to the sewage sludge. Therefore, sewage sludge contains pathogens (and heavy metals, many organic pollutants pesticides, hydrocarbons etc. if the sewage contains industrial influence). Sludge is, however, rich in nutrients such as nitrogen and phosphorous and contains valuable organic matter that is useful if soils are depleted or subject to erosion.

Options for sludge treatment include stabilisation, thickening, dewatering, drying. Sewage sludge is stabilised to reduce pathogens, to eliminate offensive odours and to inhibit, reduce or eliminate the potential for putrefaction. Moreover, stabilisation is used for volume reduction, production of usable gas (methane), and improving the dewaterability of sludge. Thickening, dewatering, drying are used to remove water from sewage sludge. Several techniques are used in dewatering devices for removing moisture. A technique close to nature and very effective is dewatering in drying beds. The principal advantages of drying beds are low costs, infrequent attention required, and high solids content in the dried product, especially in arid climates. Disadvantages are the large space required, effects of climatic changes on drying characteristics, labour-intensive sludge removal, insects and potential odours.

The purpose of sludge treatment is, besides hygienisation, to change the figures in the table above. Easily biodegradable volatile solids cause odour, that is why they should be reduced by stabilisation. A high water content ($\rightarrow$ low percentage of total solids) makes the handling difficult, causes high cost for transportation and storage and should therefore be reduced.

Sludge Pumping Systems

Sludge produced in wastewater treatment plants must be conveyed from point to point in the plant in conditions ranging from a watery sludge to a thick sludge. Sludge may also be pumped off-site for long distances for treatment. For each type of sludge and pumping application a different type of pump may be needed.

Pumps can generally be divided into centrifugal pumps (with different impellers) and displacement pumps (progressive cavity, rotary lobe, piston). Centrifugal pumps are suitable for high flow rates and low solid contents, the problem is choosing a proper size. At any given speed, centrifugal pumps operate well only if the pumping head is within a relatively narrow range. The variable nature of sludge, however, causes pumping heads to change. The selected pump must have sufficient clearance to pass the solids without clogging. Usual centrifugal pumps can cause the break up of flocculent particles in activated sludge. Good experiences in this respect have been made with screw-shaped impellers.

Table: Suitable pumps for sludge

Type of pump	Applicable for:	Advantages	Disadvantages
Centrifugal pumps	thin sludge (max. 2.5 - 3%)		
Nonclog	activated sludge	high volume, good efficiency	potential clogging (rags etc.)
Recessed impeller	sludge with solids or grit		lower efficiency
Chopper	primary sludge	reduces clogging	lower efficiency
Progressing cavity pump	thickened sludge dewatered sludge	defined flow rates acts as check valve	can run dry grit can cause high stator wear
Rotary lobe pump	thickened sludge dewatered sludge	defined flow rates acts as check valve	grit can cause high lobe wear
Piston pump	thickened sludge dewatered sludge	high pressure	discontinuous flow

Screw pump	activated sludge	good efficiency	limited height
			high capital costs
			space requirement
Screw	dried sludge		
conveyor belt	dewatered sludge	good efficiency	high capital costs
	dried sludge		space requirement

Displacement pumps can convey dewatered sludge up to 30% TS. Displacement pumps have a fixed ratio between revolutions and flow rate. Any pipeline obstruction causes damages to the pipeline or pump. Generally check valves are not necessary. For primary sludge, a grinder normally proceeds progressive cavity pumps.

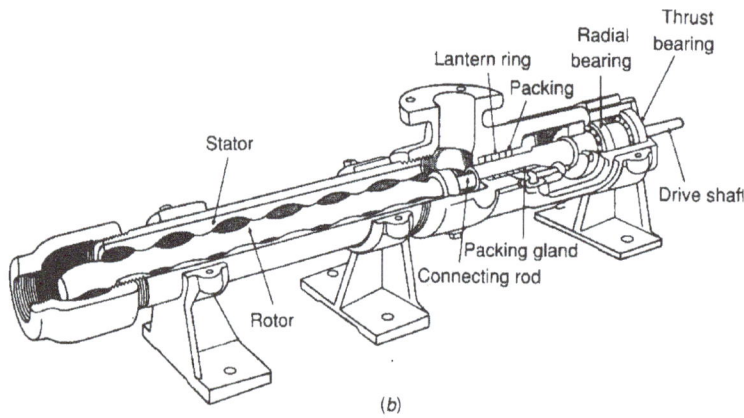

(b)

Scheme of a progressive cavity pump

Treatment Processes

Thickening

A sewage sludge thickener.

Thickening is often the first step in a sludge treatment process. Sludge from primary or secondary clarifiers may be stirred (often after addition of clarifying agents) to form larger, more rapidly settling aggregates. Primary sludge may be thickened to about 8 or 10 percent solids, while secondary sludge may be thickened to about 4 percent solids. Thickeners often resemble a clarifier with the addition of a stirring mechanism. Thickened sludge with less than ten percent solids may receive additional sludge treatment while liquid thickener overflow is returned to the sewage treatment process.

Dewatering

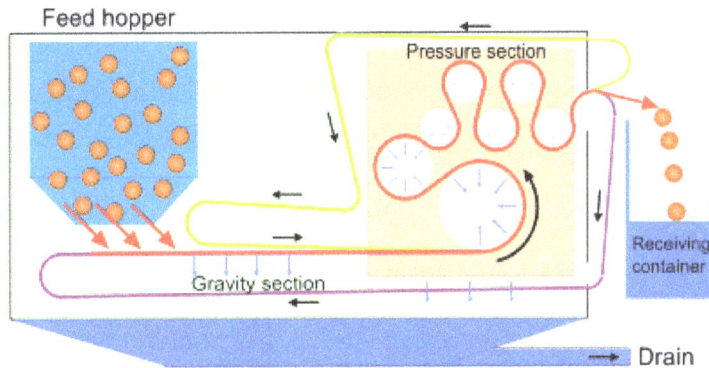

Schematic of a belt filter press to dewater sewage sludge. Filtrate is extracted initially by gravity, then by squeezing the cloth through rollers.

Water content of sludge may be reduced by centrifugation, filtration, and/or evaporation to reduce transportation costs of disposal, or to improve suitability for composting. Centrifugation may be a preliminary step to reduce sludge volume for subsequent filtration or evaporation. Filtration may occur through underdrains in a sand drying bed or as a separate mechanical process in a belt filter press. Filtrate and centrate are typically returned to the sewage treatment process. After dewatering sludge may be handled as a solid containing 50 to 75 percent water. Dewatered sludges with higher moisture content are usually handled as liquids.

Sidestream Treatment Technologies

Sludge treatment technologies that are used for thickening or dewatering of sludge have two products: the thickened or dewatered sludge, and a liquid fraction which is called sludge treatment liquids, sludge dewatering streams, liquors, centrate (if it stems from a centrifuge), filtrate (if it stems from a belt filter press) or similar. This liquid requires further treatment as it is high in nitrogen and phosphorus, particularly if the sludge has been anaerobically digested. The treatment can take place in the sewage treatment plant itself (by recycling the liquid to the start of the treatment process) or as a separate process.

Phosphorus Recovery

One method for treating sludge dewatering streams is by using a process that is also used for phosphorus recovery. Another benefit for sewage treatment plant operators of treating sludge dewatering streams for phosphorus recovery is that it reduces the formation of obstructive struvite scale in pipes, pumps and valves. Such obstructions can be a maintenance headache particularly for biological nutrient removal plants where the phosphorus content in the sewage sludge is elevated. For example, the Canadian company Ostara Nutrient Recovery Technologies is marketing a process based on controlled chemical precipitation of phosphorus in a fluidized bed reactor that recovers struvite in the form of crystalline pellets from sludge dewatering streams. The resulting crystalline product is sold to the agriculture, turf and ornamental plants sectors as fertiliser under the registered trade name "Crystal Green".

Digestion

Many sludges are treated using a variety of digestion techniques, the purpose of which is to reduce the amount of organic matter and the number of disease-causing microorganisms present in the solids. The most common treatment options include anaerobic digestion, aerobic digestion, and composting. Sludge digestion offers significant cost advantages by reducing sludge quantity by nearly 50% and providing biogas as a valuable energy source.

Anaerobic Digestion

Anaerobic digestion is a bacterial process that is carried out in the absence of oxygen. The process can either be *thermophilic* digestion, in which sludge is fermented in tanks at a temperature of 55 °C, or *mesophilic*, at a temperature of around 36 °C. Though allowing shorter retention time (and thus smaller tanks), thermophilic digestion is more expensive in terms of energy consumption for heating the sludge.

Mesophilic anaerobic digestion (MAD) is also a common method for treating sludge produced at sewage treatment plants. The sludge is fed into large tanks and held for a minimum of 12 days to allow the digestion process to perform the four stages necessary to digest the sludge. These are hydrolysis, acidogenesis, acetogenesis, and methanogenesis. In this process the complex proteins and sugars are broken down to form more simple compounds such as water, carbon dioxide, and methane.

Anaerobic digestion generates biogas with a high proportion of methane that may be used to both heat the tank and run engines or microturbines for other on-site processes. Methane generation is a key advantage of the anaerobic process. Its key disadvantage is the long time required for the process (up to 30 days) and the high capital cost. Many larger sites utilize the biogas for combined heat and power, using the cooling water from the generators to maintain the temperature of the digestion plant at the required 35 ± 3 °C. Sufficient energy can be generated in this way to produce more electricity than the machines require.

Aerobic Digestion

Aerobic digestion is a bacterial process occurring in the presence of oxygen resembling a continuation of the activated sludge process. Under aerobic conditions, bacteria rapidly consume organic matter and convert it into carbon dioxide. Once there is a lack of organic matter, bacteria die and are used as food by other bacteria. This stage of the process is known as *endogenous respiration*. Solids reduction occurs in this phase. Because the aerobic digestion occurs much faster than anaerobic digestion, the capital costs of aerobic digestion are lower. However, the operating costs are characteristically much greater for aerobic digestion because of energy used by the blowers, pumps and motors needed to add oxygen to the process. However, recent technological advances include non-electric aerated filter systems that use natural air currents for the aeration instead of electrically operated machinery.

Aerobic digestion can also be achieved by using diffuser systems or jet aerators to oxidize the sludge. Fine bubble diffusers are typically the more cost-efficient diffusion method, however, plugging is typically a problem due to sediment settling into the smaller air holes. Coarse bubble diffusers are more commonly used in activated sludge tanks or in the flocculation stages. A key component for selecting diffuser type is to ensure it will produce the required oxygen transfer rate.

Composting

Composting is an aerobic process of mixing sewage sludge with agricultural byproduct sources of carbon such as sawdust, straw or wood chips. In the presence of oxygen, bacteria digesting both the sewage sludge and the plant material generate heat to kill disease-causing microorganisms and parasites.[20] Maintenance of aerobic conditions with 10 to 15 percent oxygen requires bulking agents allowing air to circulate through the fine sludge solids. Stiff materials like corn cobs, nut shells, shredded tree-pruning waste, or bark from lumber or paper mills better separate sludge for ventilation than softer leaves and lawn clippings. Light, biologically inert bulking agents like shredded tires may be used to provide structure where small, soft plant materials are the major source of carbon.

Uniform distribution of pathogen-killing temperatures may be aided by placing an insulating blanket of previously composted sludge over aerated composting piles. Initial moisture content of the composting mixture should be about 50 percent; but temperatures may be inadequate for pathogen reduction where wet sludge or precipitation raises compost moisture content above 60 percent. Composting mixtures may be piled on concrete pads with built-in air ducts to be covered by a layer of unmixed bulking agents. Odors may be minimized by using an aerating blower drawing vacuum through the composting pile via the underlying ducts and exhausting through a filtering pile of previously composted sludge to be replaced when moisture content reaches 70 percent. Liquid accumulating in the underdrain ducting may be returned to the sewage treatment plant; and composting pads may be roofed to provide better moisture content control.

After a composting interval sufficient for pathogen reduction, composted piles may be screened to recover undigested bulking agents for re-use; and composted solids passing through the screen may be used as a soil amendment material with similar benefits to peat. The optimum initial carbon-to-nitrogen ratio of a composting mixture is between 26-30:1; but the composting ratio of agricultural byproducts may be determined by the amount required to dilute concentrations of toxic chemicals in the sludge to acceptable levels for the intended compost use. Although toxicity is low in most agricultural byproducts, suburban grass clippings may have residual herbicide levels detrimental to some agricultural uses; and freshly composted wood byproducts may contain phytotoxins inhibiting germination of seedlings until detoxified by soil fungi.

Incineration

Incineration of sludge is less common because of air emissions concerns and the supplemental fuel (typically natural gas or fuel oil) required to burn the low calorific value sludge and vaporize residual water. On a dry solids basis, the fuel value of sludge varies from about 9,500 British thermal units per pound (980 cal/g) of undigested sewage sludge to 2,500 British thermal units per pound (260 cal/g) of digested primary sludge. Stepped multiple hearth incinerators with high residence time and fluidized bed incinerators are the most common systems used to combust wastewater sludge. Co-firing in municipal waste-to-energy plants is occasionally done, this option being less expensive assuming the facilities already exist for solid waste and there is no need for auxiliary fuel. Incineration tends to maximize heavy metal concentrations in the remaining solid ash requiring disposal; but the option of returning wet scrubber effluent to the sewage treatment process may reduce air emissions by increasing concentrations of dissolved salts in sewage treatment plant effluent.

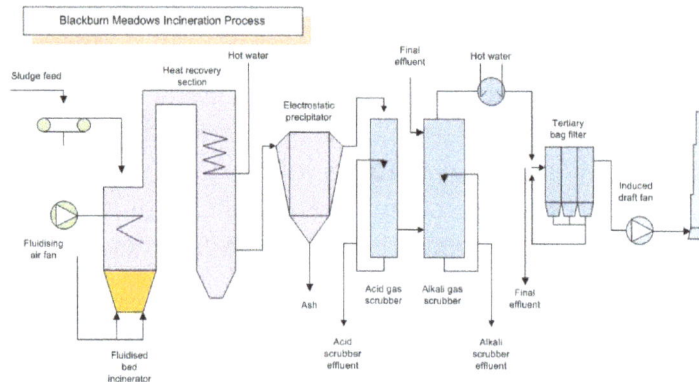

Sludge incineration process schematic (note the emphasis on air quality control).

Drying Beds

Simple sludge drying beds are used in many countries, particularly in developing countries, as they are a cheap and simple method to dry sewage sludge. Drainage water must be captured; drying beds are sometimes covered but usually left uncovered. Mechanical devices to turn over the sludge in the initial stages of the drying process are also available on the market.

Drying beds are typically composed of four layers consisting of gravel and sand. The first layer is coarse gravel that is 15 to 20 centimeters thick. Followed by fine gravel that is 10 centimeters thick. The third layer is sand that can be between 10 to 15 centimeters and serves as the filter between the sludge and gravel. Sludge dries up and water percolates to the first layer that is collected at the drainage pipe that is beneath all layers.

Sewage sludge after drying in a sludge drying bed.

This simple evaporative sludge drying bed near Damascus in Syria illustrates
the initial consistency of primary sludge being discharged from
the primary settling tank via the pipe in the foreground.

Emerging Technologies

The thermal hydrolysis system at the Blue Plains treatment plant
in Washington, D.C. is the largest in the world

- Phosphorus recovery from sewage sludge or from sludge dewatering streams is receiving increased attention particularly in Sweden, Germany and Canada, as phosphorus is a limited resource (a concept also known as "peak phosphorus") and is needed as fertilizer to feed a growing world population. Phosphorus recovery methods from wastewater or sludge can be categorized by the origin of the used matter (wastewater, sludge liquor, digested or non-digested sludge, ash) or by the type of recovery processes (precipitation, wet-chemical extraction and precipitation, thermal treatment). Research on phosphorus recovery methods from sewage sludge has been carried out in Sweden and Germany since around 2003, but the technologies currently under development are not yet cost effective, given the current price of phosphorus on the world market.

- The Omni Processor is a process that is currently under development that treats sewage sludge and can generate a surplus of electrical energy if the input materials have the right level of dryness.

- Thermal depolymerization produces light hydrocarbons from sludge heated to 250 °C and compressed to 40 MPa.

- Thermal hydrolysis is a two-stage process combining high-pressure boiling of sludge, followed by a rapid decompression. This combined action sterilizes the sludge and makes it more biodegradable, which improves digestion performance. Sterilization destroys pathogens in the sludge resulting in it exceeding the stringent requirements for land application (agriculture). Thermal hydrolysis systems are operating at sewage treatment plants in Europe, China and North America, and can generate electricity as well as high quality sludge.

Disposal or use as Fertilizer

When a liquid sludge is produced, further treatment may be required to make it suitable for final disposal. Sludges are typically thickened and/or dewatered to reduce the volumes transported off-site for disposal. Processes for reducing water content include lagooning in drying beds to produce a cake that can be applied to land or incinerated; pressing, where sludge is mechanically filtered, often through cloth screens to produce a firm cake; and centrifugation where the sludge is thickened by centrifugally separating the solid and liquid. Sludges can be disposed of by liquid injection to land or by disposal in a landfill.

There is no process which completely eliminates the need to dispose of treated sewage sludge.

Much sludge originating from commercial or industrial areas is contaminated with toxic materials that are released into the sewers from the industrial processes. Elevated concentrations of such materials may make the sludge unsuitable for agricultural use and it may then have to be incinerated or disposed of to landfill.

Examples

Edmonton, Alberta, Canada

The Edmonton Composting Facility, in Edmonton, Alberta, Canada, is the largest sewage sludge composting site in North America.

New York City, U.S.

Sewage sludge can be superheated and converted it into pelletized granules that are high in nitrogen and other organic materials. In New York City, for example, several sewage treatment plants have dewatering facilities that use large centrifuges along with the addition of chemicals such as polymer to further remove liquid from the sludge. The product which is left is called "cake," and that is picked up by companies which turn it into fertilizer pellets. This product, also called biosolid, is then sold to local farmers and turf farms as a soil amendment or fertilizer, reducing the amount of space required to dispose of sludge in landfills.

Southern California, U.S.

In the very large metropolitan areas of southern California inland communities return sewage sludge to the sewer system of communities at lower elevations to be reprocessed at a few very large treatment plants on the Pacific coast. This reduces the required size of interceptor sewers and allows local recycling of treated wastewater while retaining the economy of a single sludge processing facility and is an example of how sewage sludge can help solve an energy crisis.

Controversy

Some campaigners perceive sewage sludge treatment as a problem and a danger to the environment - largely because systems in most industrialised countries mix industrial wastes with household sewerage. This has led some to claim that the term "biosolids" was created by the sewage treatment industry in order to take the focus off the origins of the material to make reuse more acceptable to the public, and some studies have suggested that this is in fact a form of propaganda.

Coagulation Water Treatment

The coagulation process involves adding iron or aluminum salts, such as aluminum sulphate, ferric sulphate, ferric chloride or polymers, to the water. These chemicals are called coagulants, and have a positive charge. The positive charge of the coagulant neutralizes the negative charge of dissolved and suspended particles in the water. When this reaction occurs, the particles bind together, or coagulate (this process is sometimes also called flocculation). The larger particles, or floc, are heavy and quickly settle to the bottom of the water supply. This settling process is called sedimentation. The following diagram illustrates the basic reactions and processes that occur during coagulation.

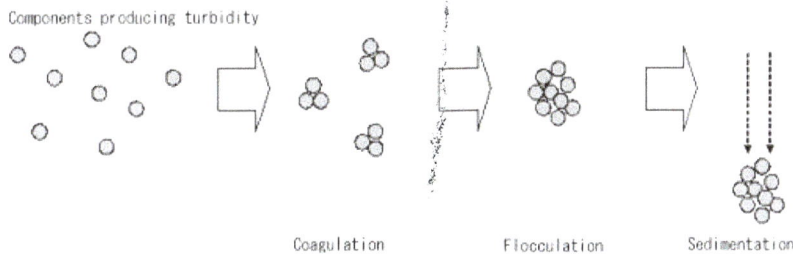

Components producing turbidity

Coagulation   Flocculation   Sedimentation

Process of Coagulation, Flocculation and Sedimentation

The chart below shows the length of time that is required for particles of different sizes to settle through the water.

Diameter of Particle	Type of Particle	Settling time through 1 m. of water
10mm	Gravel	1 seconds
1mm	Sand	10 seconds
0.1mm	Fine Sand	2 minutes
10 micron	Protozoa, Algae, Clay	2 hours
1 micron	Bacteria, Algae	8 days
0.1 micron	Viruses, Colloids	2 years
10 nm	Viruses, Colloids	20 years
1 nm	Viruses, Colloids	200 years

Settling Time for Particles of Various Diameters

In a water treatment facility, the coagulant is added to the water and it is rapidly mixed, so that the coagulant is circulated throughout the water. The coagulated water can either be filtered directly through a medium filter (such as sand and gravel), a microfiltration or ultrafiltration membrane, or it can be moved to a settling tank. In a settling tank, or clarifier, the heavy particles settle to the bottom and are removed, and the water moves on to the filtration step of the treatment process.

Coagulation can successfully remove a large amount of organic compounds, including some dissolved organic material, which is referred to as Natural Organic Matter (NOM) or Dissolved Organic Carbon (DOC). Coagulation can also remove suspended particles, including inorganic precipitates, such as iron. A large amount of DOC can give water an unpleasant taste and odour, as well as a brown discolouration. While coagulation can remove particles and some dissolved matter, the water may still contain pathogens. In an international report published in 1998, it was found that coagulation and sedimentation can only remove between 27 and 84 percent of viruses and between 32 and 87 percent of bacteria. Usually, the pathogens that are removed from the water are removed because they are attached to the dissolved substances that are removed by coagulation. In the picture below, the coagulants have been added to the water, and the particles are starting to bind together and settle to the bottom.

Coagulation jar test in a water treatment plant

As coagulation does not remove all of the viruses and bacteria in the water, it cannot produce safe drinking water. It is, however, an important primary step in the water treatment process, because coagulation removes many of the particles, such as dissolved organic carbon, that make water difficult to disinfect. Because coagulation removes some of the dissolved substances, less chlorine must be added to disinfect the water. A municipal water treatment plant can save money by using less chlorine, and the water will be safer, because trihalomethanes (THMs) are a dangerous by-product that results from the reaction of chlorine with NOM.

With accurate dosages and proper application, the residuals of the added chemicals generally do not pose a problem. Residuals are the by-products that remain in the water after substances are added and reactions occur within the water. The particular residuals depend on the coagulant that is used. If ferric sulphate is used, iron and sulphate are added to the water. If ferric chloride is used, iron and chloride are added. And if aluminum sulphate is used, aluminum and sulphate are added. The majority of municipal water treatment plants use aluminum sulphate as the coagulation chemical. Generally, water treatment facilities have the coagulation process set up so that the coagulant chemicals are removed with the floc. However, it is widely accepted that treatment facilities that use aluminum based coagulants often have higher levels of aluminum in their treated water, but not by much. In Canada, Health Canada has different guideline parameters for treatment facilities that do not use aluminum based coagulants (i.e., 0.1 mg/L) and those that do (i.e., 0.2 mg/L).

Factors

Coagulation is affected by the type of coagulant used, its dose and mass; pH and initial turbidity of the water that is being treated; and properties of the pollutants present. The effectiveness of the coagulation process is also affected by pretreatments like oxidation.

Mechanism

In a colloidal suspension, particles will settle very slowly or not at all because the colloidal particles carry surface electrical charges that mutually repel each other. A coagulant (typically a metallic salt) with the opposite charge is added to the water to overcome the repulsive charge and "destabilize" the suspension. For example, the colloidal particles are negatively charged and alum is added as a coagulant to create positively charged ions. Once the repulsive charges have been neutralized (since opposite charges attract), the van der Waals force will cause the particles to cling together (agglomerate) and form micro floc.

Determining Coagulant Dose

Jar Test

Jar test for coagulation

The dose of the coagulant to be used can be determined via the Jar Test. The jar test involves exposing same volume samples of the water to be treated to different doses of the coagulant and then simultaneously mixing the samples at a constant rapid mixing time. The microfloc formed after coagulation further undergoes flocculation and is allowed to settle. Then the turbidity of the samples are measured and the dose with the lowest turbidity can be said to be optimum.

Streaming Current Detector

An automated device for determining the coagulant dose is the Streaming Current Detector (SCD).

The SCD measures the net surface charge of the particles and shows a streaming current value of 0 when the charges are neutralized (cationic coagulants neutralize the anionic colloids). At this value (0), the coagulant dose can be said to be optimum.

Limitations

Coagulation itself results in the formation of floc but flocculation is required to help the floc further aggregate and settle. The coagulation-flocculation process itself removes only about 60%-70% of Natural Organic Matter (NOM) and thus, other processes like oxidation, filtration and sedimentation are necessary for complete raw water or wastewater treatment. Coagulant aids (polymers that bridge the colloids together) are also often used to increase the efficiency of the process.

Factors Affecting Coagulation Water Treatment

The process of coagulation of water depends on various factors like pH of the medium, temperature of water, coagulant feed concentration, coagulant dosage, type of coagulant, mass and initial turbidity. Moreover it is also depends on pre-treatment and type of pollutants present.

Effect of pH on Coagulation

pH affects on the activities of coagulants. The optimum pH for alum coagulation is 6 to 7.5 whereas 5.0 to 8.0 are for iron. If the alkalinity is lower or higher, then the floc does not form properly. As a result, more coagulant is consumed. In this case, it is beneficial to correct the pH by adding acid or base.

Temperature

Temperature is another factor for coagulation water treatment process. It is more significant at lower turbidity. In case of alum, at low temperatures aluminium hydroxide form a strongly hydrated and very stable sol. So in winter season high coagulant are consumed. When the temperature becomes below the 5^0C, then alum or ferric salts do not work properly. So it should be consider another coagulant like polyaluminum chloride (PACl).

Type of Pollutants

The salt composition of soft water and hard water are not same. Hard water contains Ca^{2+} and Mg^{2+} ions. They can alter the charge on the colloidal particles.

Optimum Dosage

It is very significant to determine the optimum dosage of a coagulant which will give the maximum clarifying effect. Insufficient amount of coagulant cannot able to destabilize properly of the colloidal particles. On the other hand higher dosage can cause excessive sludge production, corrosion and loss of money.

Type of Coagulant

All the coagulants are not suitable for all cases. Different temperatures, pH, type of medium may vary the effectiveness of the coagulant. At lower temperature the polyaluminum chloride (PACl) may be more effective than the traditional coagulants like alum or iron salt. Same way, some pH range can be beneficial to use iron salt instead of alum.

Coagulation Jar Test

You can determine optimum process condition like dosage, pH by jar test experiments. Generally, it consist several jars filled with equal volume of water. Then test for various dosage of coagulant, pH etc.

Rotating Biological Contactor

Rotating biological contactors (RBC) are a conventional aerobic biological wastewater treatment unit. Conventional biological treatment means activated sludge systems and fixed film systems such as trickling filters, or RBC (NOLDE 1996). The advantage of all these systems is that they are compact (i.e. in densely populated urban settings) and that they efficiently reduce organic matter (JENSSEN 2004). However, they are high-tech and generally require skilled staff for construction as well as for operation. RBC can treat domestic black- or greywater and any other low- or high-strength biodegradable wastewater (e.g. industrial wastewater from food processors or paper mills). They have been found to be particularly effective for decentralised applications (on the level of a small to medium community or industry/institution), where electricity and skilled staff are available.

Treatment Process and Basic Design Principles

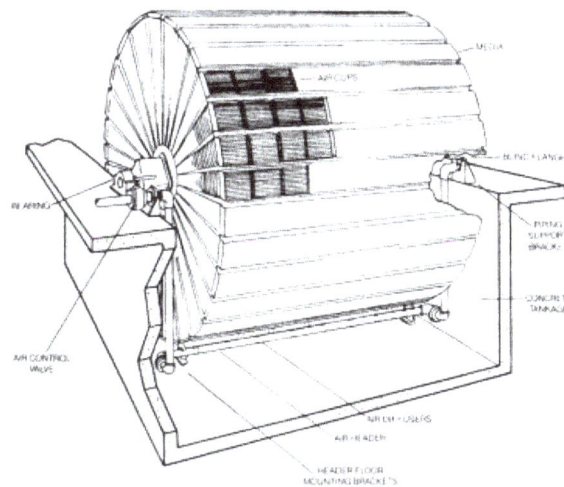

The disc is made out of light-weighted material such and usually ridged,
corrugated, or lattice-like to make as much as surface available for
the biofilm to attach

A series of circular lightweight rotating discs are mounted on a shaft through which wastewater flows. The partially submerged discs rotate through the wastewater slowly. The disks are most commonly made of high-density plastic sheets (e.g. Polyethylene, polystyrene or polyvinylchloride) and are usually ridged, corrugated, or lattice-like to increase the specific surface area (NSFC 2004). The surface of the disks provides an attachment site for bacteria and as the discs rotate, a film of biomass grows on their surfaces (NSFC 2004; WSP 2008). This biofilm is alternately exposed to either the air or the wastewater as it rotates. The oxygen necessary for the growth of these microorganisms is obtained by adsorption from the air as the biofilm on the disk is rotated out of the liquid (CRITES & TCHOBANOGLOUS 1998; SANIMAS 2005). As the biofilm passes through the liquid phase, nutrients and organic pollutants are taken up. All oxygen, nutrients and organic pollutants are necessary for the growth of the microorganism and the conversion of the organic matter to CO_2. Nitrogen is removed by nitrification and subsequent denitrification transforming it to gaseous N_2, which is released to the air. The process is optimised by adjusting the speed of rotation and the depth of submergence (METCALF & EDDY Inc. 2003). In some designs, air is added to the bottom of the tank to provide additional oxygen in case of high-strength influents (CRITES & TCHOBANOGLOUS 1998).

The submerging level varies from 40 to 80 % (CRITES & TCHOBANOGLOUS 1998) and a usual rotating speed is 1 to 2 rpm (U.S.EPA 1980). The common disc diameter is between 0.6 and 3 m (SANIMAS 2005). The degradation process is similar to the one in a trickling filter with a high rate of recirculation (CRITES & TCHOBANOGLOUS 1998). The higher contact time in RBCs due to rotation allows up to 8 to 10 times higher levels of treatment than in trickling filters (WSP 2008). Also because the rotation allows both optimum wetting and oxygen supply, RBCs are generally more reliable than other fixed-film processes. Additionally, the disc design is made in such a way that large amounts of biofilm can attach, which means that there is a large amount of biological mass present to degrade the pollutants (WSP 2007). The large amount of biomass and the stability of contact also results in an improved stability and a reduced susceptibility to changes in hydraulic or organic loading compared to conventional activated sludge processes (WSP 2007). As for all fixed-film processes, primary settling and/or screening is required for the removal of grit, debris, and excessive oil (U.S. EPA 1980, WSP 2008). Such primary treatments are typically septic tanks, Imhoff tanks or anaerobic reactors. To remove sloughing sludge, a post-settling unit (i.e. a clarifier) is also required.

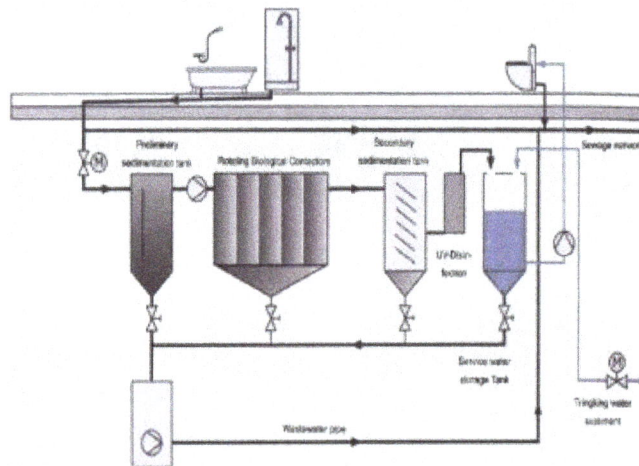

RBCs are a secondary treatment and as for all fixed-film processes, primary settling as well as sedimentation of sloughed sludge in a tertiary clarifier is required (example: greywater treatment in Germany)

The performance of RBC systems depends on the design, the temperature, the concentration of the pollutants, the rotating velocity and the hydraulic retention time. RBCs can achieve biological oxygen demand (BOD) reductions of 80 to 90 % (SANIMAS 2005; WSP 2007; WSP 2008). The removal of nitrogen (which is mostly present as ammonia) by nitrification and subsequent denitrification is also high, because both aerobic nitrifying bacteria and anaerobic denitrifying bacteria can simultaneously live in the attached biofilm (HOCHHEIMER 1998), depending on weather they are situated on the bottom of the film, close to the disc support (and thus in anaerobic or anoxic conditions) or at the top of the film exposed to the air.

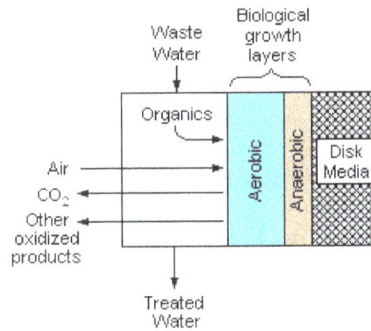

Both aerobic and anaerobic microorganisms can live in the biofilm
and contribute to the removal of pollutant form the water.

Some other microorganisms which can transform ammonia (NH3) in one single step to gaseous N2 under anaerobic conditions have also been discovered in biofilms growing on RBC. These bacteria were called annamox and resulted in the development of innovative aerobic ammonia removal and wastewater treatment processes.

Little is known about the removal of phosphorus in RBCs, but it can be presumed that large parts of the phosphorus present is either accumulated in the biofilm or in the settled and collected sludge.

RBCs can be arranged in a variety of ways depending on specific effluent characteristics and the secondary clarifier design (e.g. specifically for BOD removal or nitrification, NSFC 2004).

Excess biomass sloughs off the discs by the shearing forces exerted as the discs rotate, combined with the force of gravity (WSP 2008). The rotation movement helps to keep sloughed solids in suspension so they can be carried to a clarifier (gravity settler) for secondary settling. The collected sludge in the clarifier requires further treatment (WSP 2008) for stabilisation, such as anaerobic digestion, composting, constructed wetlands, ponds or drying. Very often in small installations, accumulated sludge is also directed back to the septic tank for storage and partial digestion (U.S.EPA 1980).

Effluents from RBC do not contain high levels of nutrients and are therefore not particularly interesting for agriculture, although they constitute a source of water. However, due to reduced removal of microorganisms (1 to 2 log units, U.S. EPA 2002), RBC effluents require a further treatment, such as sand filtration, constructed wetlands or another form of disinfection (e.g. chemical disinfection or UV disinfection).

RBCs are usually designed on the basis of hydraulic and organic loadings derived from pilot plants and other full-scale installation (WSP 2007). Hydraulic retention times (HRTs) generally lye within some hours up to two days.

Even though RBCs are resistant to shock loading, long-term high organic loading may cause anaerobic conditions, resulting in odour and poor treatment performance (U.S.EPA 1980).

Example of an underground RBC for the decentralised treatment of domestic blackwater.

Recirculation is not normally practised in package fixed-film systems since it adds to the degree of complexity and is energy and maintenance intensive. However, recirculation may be desirable in certain applications where minimum wetting rates are required for optimal performance (U.S.EPA 1980).

Units may be installed at or below ground depending upon site topography and other adjacent treatment processes. Access to all moving parts and controls is required, and proper venting of the units is paramount, especially if natural ventilation is being used to supply oxygen (U.S.EPA 1980).

RBCs are often covered with a fibreglass housing to protect the disks from sunlight, wind, rain and low temperatures as performance of RCS drops considerably at air temperatures below 12°C (U.S.EPA 2002; NSFC 2004).

Costs Considerations

Observed costs for RBCs are highly variable depending on climate and location. Generally, RBCs involve high capital costs as not all materials may be locally available and motor and special material for rotation is required. Another cost factor may be manufacture and implementation, which requires skilled experts (SANIMAS 2005).

Operation and maintenance costs are also relatively high, because operation requires a continuous electricity supply and supervision requires semi-skilled labour (U.S.EPA 1980) and professional operator (SANIMAS 2005).

Operation and Maintenance

Large-scale RBCs are often covered to protect them form cold temperatures, rain, wind and sun. Sometimes, artificial aeration is required to keep the process aerobic when the systems are covered

During operation, the system must be supervised by professional operators (SANIMAS 2005). Maintenance includes lubrication of moving parts, motors and bearings; replacing seals, motors, servicing bearings; and cleaning the attached-growth media (spray-washing of discs and purging of settled sludge) (METCALF & EDDY Inc. 2003; WSP 2007). The discs may be also checked for debris accumulation, ponding and excessive or not sufficient biomass accumulation (U.S.EPA 2002).

Although fixed film units such as RBC and trickling filters are operation- and maintenance-intensive, they do not require seeding with bacterial cultures (as do anaerobic processes such as anaerobic baffled reactors, septic tanks, upflow anaerobic sludge blanket reactors or anaerobic digesters) and the start-up phase is therefore considerably shorter. However, it takes 6 to 12 weeks for the biofilm to establish for a good treatment performance (U.S.EPA 2002).

Health Aspects

For correct operation, RBCs are covered and thus protected from contact with humans or animals. However, excess sludge as well as the effluent require post-treatment to remove pathogenic microorganisms.

In any case (i.e. for operation and maintenance) direct contact with the biomass growing on the discs, the effluent or the sludge should be avoided.

For discharge or reuse, a treatment unit allowing further pathogen removal should be considered.

Applicability

RBCs can achieve a high removal of biodegradable organic pollutants form domestic black- or greywater as well as from high-strength industrial wastewater (e.g. from dairies, bakeries, food processors, pulp, paper mills, WSP 2008). A great variety of applications are known, either as post-treatment for activated sludge processes in conventional domestic wastewater treatment plants, or for decentralized application at the level of small to medium-sized communities, industries or institutions (WSP 2007; WSP 2008).

They are adapted for urban areas mostly: land requirements are low, but continuous and consequent energy supply as well as semi-skilled labour are indispensable.

Some of the material may be locally available, however, the system can only be planned and implemented by experts (SANIMAS 2005).

Bar Screen

Bar screen and screening water treatment is the first process unit operation used at wastewater treatment plants. Screening removes objects such as rags, paper, plastics and metals to prevent damage and clogging of downstream equipment and piping.Cleaning frequency depends on the characteristics of the wastewater entering a plant.

Bar Screen Operation and Maintenance Considerations

- Check and clean the bar screen at frequent intervals.

- Do not allow solids to overflow /escape from the screen bar screen.

- Ensure no large gaps are formed due to the breakage of the screening water treatment.

- Replace breakage bar screen immediately.

- Mechanically cleaned screening system to remove larger materials because they reduce labor cost and they improve flow conditions and screening capture.

- Mechanically cleaned bar screen should have a standby screen to put in operation when the primary screening device is out of service.

Trouble Shooting of Screening Water Treatment

S.NO	PROBLEM	CAUSE
1	Large particles pass through and check the pumps	Poor operation /screen damaged
2	Up stream water levels is much higher than down stream level	Poor operation (inadequate cleaning)
3	Excessive collection of trash on screening water treatment	Poor operation
4	Excessive odor	Poor operation /trash disposal practices.

Daily and Weekly Maintenance Work

- Check gear box oil qty periodically and completely drain out oil and replace afresh as per manufacture's recommendation.

- Check oil pump every day ,top up if necessary.

- Check every day chain alignments and must periodically 4 hrs once chain cleaned or removed impurities materials.

Main use of Screening Water Treatment

- To remove the suspended solids.

- To avoided the pump cloaking.

- To increase the Bacteria attachment in the FBBR system.

- To reduce the Sludge Quantity in filter press.

- To reduce the short-circuit in Electro Coagulation System.

- To avoided the solids cloaking in the air distribution system in Equalization tank.

- To increase the air volume in Equalization tank.

Project Brief of Screening Water Treatment Plant

Auto is engaged into printing of textile garments using pigments and reactive colours. The Effluent Treatment Plant is designed to treat 40m3 per day of the effluent to achieve the norms prescribed by the local governing bodies and BSR guidelines. The Factory presently produce only up to 10m3/Day Effluent. The ETP is designed for higher capacity considering the future expansion and addition of Washing Process.

Plan Capacity of Screening Water Treatment Plant

The Effluent screening water treatment Plant is designed to treat wastewater generated from the processing unit. The Effluent treatment plant is designed to treat 40m3/day of Washing & Printing combined effluent generated from the process house (Washing 15m3/Day & Printing 5m3/Day). The plant will be capable of operating at the flow rate of 2m3/hr. The Present Effluent generation is 10m3/Day and the Plant is operated for 10-12 Hours per Day with 50% flow rate.

Organisms Involved in Water Purification

Water purification using organisms is a great natural method to treat mainly used water.

The implementing of purification using organisms is known and works in practice for many years now, the principals of the method are natural and occur in nature all the time regardless of human intervention. But the use of this type of biological water treatment method is becoming more known and also more common due to the overall understanding that mankind needs to find more sustainable and ecological ways to live, treat our waste and co-exist with other systems on the planet.

The major microbial populations found in wastewater treatment systems are bacteria, protozoa, viruses, fungi, algae and helminthes. The presence of most of these organisms in water leads to spread of diseases[1]. The majority of waterborne microorganisms that cause human disease come from animal and human fecal wastes. These contain a wide variety of viruses, bacteria, and protozoa that may get washed into drinking water supplies or receiving water bodies[2]. Microbial pathogens are considered to be critical factors contributing to numerous waterborne outbreaks. Many microbial pathogens in wastewater can cause chronic diseases with costly long-term effects, such as degenerative heart disease and stomach ulcer. The density and diversity of these pollutants can vary depending on the intensity and prevalence of infection in the sewered community. The detection, isolation and identification of the different types of microbial pollutants in wastewater are always difficult, expensive and time consuming. To avoid this, indicator organisms are always used to determine the relative risk of the possible presence of a particular pathogen in wastewater.

Viruses are among the most important and potentially most hazardous pollutants in wastewater. They are generally more resistant to treatment, more infectious, more difficult to detect and require smaller doses to cause infections[4,5]. Because of the difficulty in detecting viruses, due to their low numbers, bacterial viruses (bacteriophages) have been examined for use in faecal pollution and the effectiveness of treatment processes to remove enteric viruses. Bacteria are the most common microbial pollutants in wastewater. They cause a wide range of infections, such as diarrhea, dysentery, skin and tissue infections, etc. Disease-causing bacteria found in water include different types of bacteria, such as *E. coli* O157:H7; *Listeria* sp., *Salmonella* sp., *Leptosporosis* sp., etc (CDC, 1997). The major pathogenic protozoans associated with wastewater are *Giardia* sp. and *Cryptosporidium* sp. They are more prevalent in wastewater than in any other environmental source.

Pollutants in Wastewater

Pathogens

Parasites, bacteria and viruses may be injurious to the health of people or livestock ingesting the polluted water. These pathogens may have originated from sewage or from domestic or wild bird or mammal feces. Pathogens may be killed by ingestion by larger organisms, oxidation, infection by phages or irradiation by ultraviolet sunlight unless that sunlight is blocked by plants or suspended solids.

Suspended Solids

Particles of soil or organic matter may be suspended in the water. Such materials may give the water a cloudy or turbid appearance. The anoxic decomposition of some organic materials may give rise to obnoxious or unpleasant smells as sulphur containing compounds are released.

Nutrients

Compounds containing nitrogen, potassium or phosphorus may encourage growth of aquatic plants and thus increase the available energy in the local food-web. this can lead to increased concentrations of suspended organic material. In some cases specific micro-nutrients may be required to allow the available nutrients to be fully utilised by living organisms. In other cases, the presence of specific chemical species may produce toxic effects limiting growth and abundance of living matter.

Metals

Many dissolved or suspended metal salts exert harmful effects in the environment sometimes at very low concentrations. Some aquatic plants are able to remove very low metal concentrations, with the metals ending up bound to clay or other mineral particles.

Roles and Dynamics of Microorganisms in Wastewater Treatment Systems

The major microbial populations found in wastewater treatment systems are bacteria, protozoa, viruses, fungi, algae and helminthes. The presence of most of these organisms in water leads to spread of diseases.

Bacteria

In wastewater treatment systems, bacteria play vital role in the conversion of organic matter present to less complex compounds. In terms of size, bacteria range from 0.2- 2.0 μm in diameter and are responsible for most of the wastewater treatment in septic tanks. Though not all bacteria are harmful, a number of them cause water-related diseases in human and animals. Some of these diseases include cholera, dysentery, typhoid fever, salmonellosis and gastroenteritis. Bacteria can be found entangled in flocs as in the case of activated sludge, where some play important role in biological treatment while some like the filamentous bacteria can cause serious problem in settling and foaming. Waterborne gastroenteritis of unknown cause is frequently reported, with susceptible agent being bacteria. Certain strains of *Pseudomonas* and *Escherichia coli* which may affect the newborn are potential sources of this disease. These strains of microbes have also been implicated in gastrointestinal disease outbreaks.

Bacteria are of the greatest numerical importance in wastewater treatment systems. The majority is facultative living in either the presence or absence of oxygen. Although both heterotrophic and autotrophic bacteria are found in wastewater treatment systems thepredominant ones are the heterotrophic bacteria. Generally, heterotrophic bacteria obtain their energy from the carbonaceous organic matter in wastewater effluent. The energy obtained is used for the synthesis of new cells and also for the release of energy through the conversion of organic matter and water. Some important bacteria genera that are found in wastewater treatment systems are *Achromobacter, Alcaligenes, Arthrobacter, Citromonas, Flavobacterium, Pseudomonas, Zoogloe* and *Acinetobacter*.

In wastewater treatment systems, bacteria are responsible for the stabilization of influent wastes. The majority of the bacteria are known to form floc particles. The floc particles are clusters of bacteria that break down waste. Also, the floc particles also serve as sites on which waste can be absorbed and broken down. Filamentous bacteria form, which trichomes or filaments provide a backbone for the floc particles, allowing the particles to grow in size and withstand the shearing action in the treatment process. When filamentous bacteria are present in excessive numbers or length, they often cause solid/liquid separation or settleability problems.

Furthermore, bacteria are the most common microbial pollutants in wastewater. The presence of pathogenic bacteria can be indicated using the tests for total and faecal coliforms. Conventionally, the detection of feacal coliform is generally accepted as a reliableindicator of faecal contamination. *Escherichia coli*is also regarded as agood and reliable indicator forfaecal pollution from animal and human sources since it is known not to last for long periods outside the faecal environment. The tests for total and faecal coliforms can be carried out, using either the traditional or enzymatic methods. Traditionally, the tests for total and faecal coliforms are carried out using the multiple-tube fermentation or by membrane filtration techniques. While the multiple-tube fermentation technique is used for medium or highly contaminated waters, the membrane filtration technique is used for low or very low contaminated waters.

Protozoa

Protozoa are microscopic, unicellular organisms that are also found in the wastewater treatment systems. They perform many beneficial functions in the treatment process, including the clarification of the secondary effluent through the removal of bacteria, flocculation of suspended material and as bioindicators of the health of the sludge. The protozoa that inhabit the wastewater treatment systems are capable of movement in at least one stage of their development. They are 10 times bigger than bacteria; they are unicellular organisms with membrane enclosed organelles.

Protozoa prey on pathogenic bacteria which make it have an advantage in wastewater. They can be classified into five groups depending on their mode of locomotion, which are the free swimming ciliates, crawling ciliates, stalked and sessile ciliates, flagellates and amoeboid.

In wastewater treatment systems, protozoa are useful biological indicators of the condition of the systems. Although some protozoa are able to survive up to 12 h in the absence of oxygen, they are generally known as obligate aerobes, hence prove to be excellent indicators of an aerobic environment. Additionally, they serve as indicators of a toxic environment and are capable of exhibiting greater sensitivity to toxicity than bacteria. An indication of possible toxicity in a treatment system is the absence of or a lack of mobility of protozoa. The presence of large numbers of highly evolved protozoa in the biological mass in a wastewater treatment system is indicated as a hallmark of a well-operated and stable system.

They can be placed into one of five groups, according to their means of locomotion. These groups are the free-swimming ciliates, crawling ciliates and stalked/sessile ciliates, flagellates and amoebae. The three types of ciliates are free-swimming ciliates, crawling ciliates and stalked ciliates. All of these three have short hair-like structures or cilia that beat in unison to produce water current for locomotion and capturing bacteria. The water current moves suspended bacteria into a mouth opening.

In the aeration tank of biological processes, a true trophic web is established. The biological system of these plants consists of populations in continuous competition with each other for food. The growth of decomposers, prevalently heterotrophic bacteria, depends on the quality and quantity of dissolved organic matter in the mixed liquor. For predators, on the other hand, growth depends on the available prey. Dispersed bacteria are thus food for heterotrophic flagellates and bacterivorous ciliates, which, in turn, become the prey of carnivorous organisms. The relationships of competition and predation create oscillations and successions of populations until dynamic stability is reached. This is strictly dependent on plant management choices based on design characteristics aimed at guaranteeing optimum efficiency.

Some ciliates, however, are predators of other ciliates or are omnivorous, feeding on a variety of organisms including small ciliates, flagellates and dispersed bacteria. All bacterivorous ciliates rely on ciliary currents to force suspended bacteria to the oral region. Ciliated protozoa are numerically the most common species of protozoa in activated sludge, but flagellated protozoa and amoebae may also be present. The species of ciliated protozoa most commonly observed in wastewater treatment processes include *Aspidiscacostata, Carchesiumpolypinum, Chilodonellauncinata, Operculariacoarcta, Operculariamicrodiscum, Trachelophyllumpusillum, Vorticella convallaria* and *Vorticella microstoma*.

There is indication that the free-swimming ciliates such as *Litonotus* sp. and *Paramecium* sp., which have cilia on all their body surfaces are typically found suspended or swimming freely in the bulk solution. On the other hand, the crawling ciliates, such as *Aspidisca* sp. and *Euplotes* sp. possess cilia only on their ventral or belly surface where the mouth opening is located. The crawling ciliates are usually found on floc particles while the stalked ciliates, such as *Carchesium* sp. and *Vorticella* sp., possess their cilia around the mouth opening only and are attached to floc particles. They have enlarged anterior portion and a slender posterior portion. The beating of the cilia and the springing action of the stalk produce a water vortex that draws dispersed bacteria into the mouth opening.

In wastewater systems, two types of amoebae are predominant, naked, such as *Actinophyrs* sp., *Mayorella* sp. and *Thecamoeba* sp. and the shelled amoebae or testate amoebae, e.g. *Cyclopyxis* sp. The naked amoeba lack any protective covering while shelled amoebae possess a protective covering that consists of calcified material. flagellated protozoa are oval in shape and possess one or more whip-like flagella. In wastewater treatment systems, the flagellated protozoa are propelled through the system by the help of the flagella in a cork-screw pattern of locomotion.

Viruses

Viruses are also found in wastewaters, particularly human viruses that are excreted in large quantities in faeces. Viruses that are native to animals and plants exist in smaller quantities in wastewater, although bacterial viruses may also be present. They are the causative agents of some water-related infections in humans, such as gastrointestinal and respiratory infections, conjunctivitis and meningitis. It is reported that a majority of waterborne diseases due to unidentified sources were caused by enteric viruses. They are very notorious and persistent when present in wastewater and can remain a viable source of infection for months after their entry into the wastewater.

Fungi

Fungi are also part of the microorganisms found in wastewater treatment systems. Fungi are multicellular organisms that are also constituents of the activated sludge. Under certain environmental conditions in a mixed culture, they metabolize organic compounds and can successfully compete with bacteria. Also, a small number of fungi are capable of oxidizing ammonia to nitrite, and fewer still to nitrate. The most common sewage fungus organisms are *Sphaerotilus natans* and *Zoogloea* sp.

A number of filamentous fungi are found naturally in wastewater treatment systems as spores or vegetative cells, although they can also metabolize organic substances. A number of fungi species, such as *Aspergillus, Penicillium, Fusarium, Absidia*and a host of others have been implicated in the removal of carbon and nutrient sources in wastewater. Some fungi are also reported to have the ability to breakdown organic matter present in the sludge system. In a system with low pH, where bacterial growth is inhibited, the main role of the fungi is the breakdown down of organic matter. Additionally, some fungi use their fungal hyphae for trapping and adsorbing suspended solids to accomplish their energy and nutrient requirements. Some filamentous fungi have been reported to secrete some enzymes which help in the degradation of substrates during wastewater treatment.

Algae

Algae can be found in wastewater because they are able to use solar energy for photosynthesis as well as nitrogen and phosphorus for their growth leading to eutrophication. Some types of algae that can be found in wastewater include *Euglena* sp., *Chlamydomonas* sp., and *Oscillatoria* sp. Algae are significant organisms for biological purification of wastewater because they can be able to accumulate plant nutrients, heavy metals, and pesticides, organic and inorganic toxic substances. The use of microalgae in biological wastewater treatment has gained a lot of importance over the years.

High rate algal pond is shallow and equipped with mechanical aeration and mixing by means of paddle wheels, 90% of BOD and 80% of nitrogen and phosphorus are treated in high rate algal ponds. Construction and energy costs are highly lower and the land requirement is not up to that of facultative pond in constructed wetlands.

Helminth

Nematodes are aquatic animals present in fresh, brackish and salt waters and wet or humid soil worldwide. Freshwater nematodes can be present in sand filters and aerobic treatment plants. They are present in large numbers in secondary wastewater effluents, biofilters and biological contractors. Freshwater nematodes inhabit freshwater below the water table with species utilizing oxygen dissolved in the fresh water. Nematodes are part of the ecosystem, serving as food for small invertebrates. They crawl onto floc particles and move in whip-like fashion when in the free-living mode. They secrete a sticky substance to be able to anchor to a substrate (media), so that anchored nematodes can feed without interference from currents or turbulence. A lack of nematode activity can be one of the bio-indicators of a toxic condition that may be developing in the treatment process.

There are many species of parasitic worms that have human hosts. Some of them have the ability to cause serious illnesses. They can be grouped into three, the roundworms, the flatworms and the annelids. The flatworms are divided into two groups, the tapeworms, which posses segments on their body and the flukes, which have a single, flat and unsegmented body. A common characteristic of most helminthes is that they reproduce through eggs, although the eggs may differ in size and shape. In wastewater, most helminthes eggs are not always infectious. For the eggs to be infectious they need to be viable and larval development must occur. The larval development occurs at required levels of temperature and moisture. Helminth eggs can remain viable for 1-2 months in crops and it takes longer months in soil, fresh water and sewage and can take years in faeces and sludge. Helminthes eggs are inactivated at elevate temperatureised (above 40°C) and reduced moisture (below 5%), although these conditions are not readily achieved during wastewater treatment; In typical water treatment, helminthes eggs are removed by physical means, such as sedimentation, filtration or coagulation-flocculation.

Secondary Treatment

Secondary treatment of sewage and other wastewater is the stage of wastewater treatment designed to substantially degrade the biological content of the sewage. This usually uses biological processes. Municipal and industrial plants usually use aerobic biological processes.

Suspended processes, particularly activated sludge, are most common in medium to large-scale plants; fixed film methods such as roughing filters need less maintenance and control, and are more resilient, and are appropriate where cost and maintenance are major issues.

Anaerobic treatment is sometimes used, in the form of septic tanks and in biogas digesters. in the case of septic tanks the primary and secondary phases are combined in one unit. If biogas digesters are used for secondary treatment, the primary treatment phase is reduced or emitted (aiming to remove matter such as gravel and litter rather than sewage solids).

Brief Explanation

Once wastewater has gone through the Primary treatment stage the Effluent will undergo a secondary treatment in order to remove both small suspended solids and BOD_5 (five day biochemical oxygen demand) that pass through the primary treatment stage. All secondary treatment systems use a biological process to break down organic matter. Microorganisms are introduced to the wastewater and consume the organic matter, oxygen is delivered to the system ensuring microorganism survival. Oxygen delivery differs among the various systems. This biological process occurs naturally in nature, but is accelerated in secondary treatment systems. Typically 85% of BOD and suspended solids are removed during this process. Water exiting secondary treatment will still carry nitrogen, phosphorus, heavy metals, Pathogens, and bacteria. For further removal of pollutants the water is transported to a tertiary treatment system and disinfection. There are a variety of secondary treatment processes; the following are conventional processes used by treatment plants:

- Activated sludge

- Trickling filter

- Non-electric secondary filtration (FilterPod)

- Oxidation ponds

There are pros and cons to each of these three processes. Operational and initial costs along with space are three factors that will often determine which technique is appropriate. Space is influenced by population size and cost of land. For example, oxidation ponds require large areas of land if land is costly or needed for housing oxidation ponds are not a likely option. Additionally wastewater treatment plants need to consider maintenance, reliability, and effectiveness of the system.

Process Upsets

Process upsets are temporary decreases in treatment plant performance caused by significant population change within the secondary treatment ecosystem. Conditions likely to create upsets include for example toxic chemicals and unusually high or low concentrations of organic waste BOD providing food for the bioreactor ecosystem.

Toxicity

Waste containing biocide concentrations exceeding the secondary treatment ecosystem tolerance level may kill a major fraction of one or more important ecosystem species. BOD reduction normally accomplished by that species temporarily ceases until other species reach a suitable population to utilize that food source, or the original population recovers as biocide concentrations decline.

Dilution

Waste containing unusually low BOD concentrations may fail to sustain the secondary treatment population required for normal waste concentrations. The reduced population surviving the starvation event may be unable to completely utilize available BOD when waste loads return to normal. Dilution may be caused by addition of large volumes of relatively uncontaminated water such as stormwater runoff into a combined sewer. Smaller sewage treatment plants may experience dilution from cooling water discharges, major plumbing leaks, firefighting, or draining large swimming pools.

A similar problem occurs as BOD concentrations drop when low flow increases waste residence time within the secondary treatment bioreactor. Secondary treatment ecosystems of college communities acclimated to waste loading fluctuations from student work/sleep cycles may have difficulty surviving school vacations. Secondary treatment systems accustomed to routine production cycles of industrial facilities may have difficulty surviving industrial plant shutdown. Populations of species feeding on incoming waste initially decline as concentration of those food sources decrease. Population decline continues as ecosystem predator populations compete for a declining population of lower trophic level organisms.

Peak Waste Load

High BOD concentrations initially exceed the ability of the secondary treatment ecosystem to utilize available food. Ecosystem populations of aerobic organisms increase until oxygen transfer limitations of the secondary treatment bioreactor are reached. Secondary treatment ecosystem populations may shift toward species with lower oxygen requirements, but failure of those species to use some food sources may produce higher effluent BOD concentrations. More extreme increases in BOD concentrations may drop oxygen concentrations before the secondary treatment ecosystem population can adjust, and cause an abrupt population decrease among important species. Normal BOD removal efficiency will not be restored until populations of aerobic species recover after oxygen concentrations rise to normal.

Design for Damage Control

Measures creating uniform wastewater loadings tend to reduce the probability of upsets. Fixed-film or attached growth secondary treatment bioreactors are similar to a plug flow reactor model circulating water over surfaces colonized by biofilm, while suspended-growth bioreactors resemble a continuous stirred-tank reactor keeping microorganisms suspended while water is being treated. Secondary treatment bioreactors may be followed by a physical phase separation to remove biological solids from the treated water. Upset duration of fixed film secondary treatment systems may be longer because of the time required to recolonize the treatment surfaces. Suspended growth ecosystems may be restored from a population reservoir. Activated sludge recycle systems provide an integrated reservoir if upset conditions are detected in time for corrective action. Sludge recycle may be temporarily turned off to prevent sludge washout during peak storm flows when dilution keeps BOD concentrations low. Suspended growth activated sludge systems can be operated in a smaller space than fixed-film trickling filter systems that treat the same amount of water; but fixed-film systems are better able to cope with drastic changes in the amount of biological material and can provide higher removal rates for organic material and suspended solids than suspended growth systems.

Wastewater flow variations may be reduced by limiting stormwater collection by the sewer system, and by requiring industrial facilities to discharge batch process wastes to the sewer over a time interval rather than immediately after creation. Discharge of appropriate organic industrial wastes may be timed to sustain the secondary treatment ecosystem through periods of low residential waste flow. Sewage treatment systems experiencing holiday waste load fluctuations may provide alternative food to sustain secondary treatment ecosystems through periods of reduced use. Small facilities may prepare a solution of soluble sugars. Others may find compatible agricultural wastes, or offer disposal incentives to septic tank pumpers during low use periods.

Process Types

Filter Beds (Oxidizing Beds)

In older plants and those receiving variable loadings, trickling filter beds are used where the settled sewage liquor is spread onto the surface of a bed made up of coke (carbonized coal), limestone chips or specially fabricated plastic media. Such media must have large surface areas to support the biofilms that form. The liquor is typically distributed through perforated spray arms. The distributed liquor trickles through the bed and is collected in drains at the base. These drains also provide a source of air which percolates up through the bed, keeping it aerobic. Bio-films of bacteria, protozoa and fungi form on the media's surfaces and eat or otherwise reduce the organic content. The filter removes a small percentage of the suspended organic matter, while the majority of the organic matter supports microorganism reproduction and cell growth from the biological oxidation and nitrification taking place in the filter. With this aerobic oxidation and nitrification, the organic solids are converted into biofilm grazed by insect larvae, snails, and worms which help maintain an optimal thickness. Overloading of beds may increase biofilm thickness leading to anaerobic conditions and possible bioclogging of the filter media and ponding on the surface.

Rotating Biological Contactors

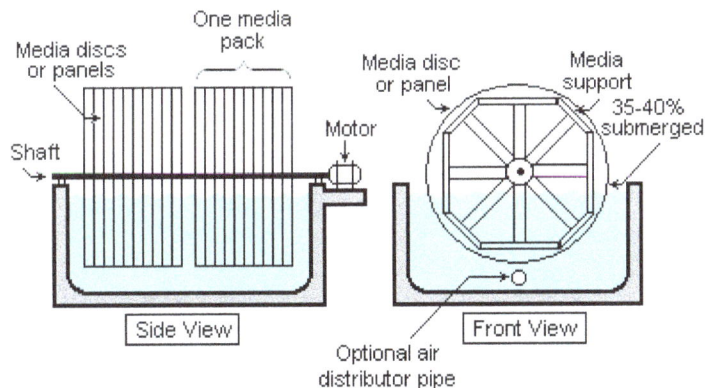

Schematic of a typical rotating biological contactor (RBC). The treated effluent clarifier/settler is not included in the diagram.

Rotating biological contactors (RBCs) are robust mechanical fixed-film secondary treatment systems capable of withstanding surges in organic load. RBCs were first installed in Germany in 1960 and have since been developed and refined into a reliable operating unit. The rotating disks support the growth of bacteria and micro-organisms present in the sewage, which break down and stabilize organic pollutants. To be successful, micro-organisms need both oxygen to live and food to grow. Oxygen is obtained from the atmosphere as the disks rotate. As the micro-organisms grow, they build up on the media until they are sloughed off due to shear forces provided by the rotating discs in the sewage. Effluent from the RBC is then passed through a secondary clarifier where the sloughed biological solids in suspension settle as a sludge.

Activated Sludge

Activated sludge is a common suspended-growth method of secondary treatment. Activated sludge plants encompass a variety of mechanisms and processes using dissolved oxygen to promote growth of biological floc that substantially removes organic material.[12-13] Biological floc is an ecosystem of living biota subsisting on nutrients from the inflowing primary clarifier effluent. These mostly carbonaceous dissolved solids undergo aeration to be broken down and either biologically oxidized to carbon dioxide or converted to additional biological floc of reproducing micro-organisms. Nitrogenous dissolved solids (amino acids, ammonia, etc.) are similarly converted to biological floc or oxidized by the floc to nitrites, nitrates, and, in some processes, to nitrogen gas through denitrification. While denitrification is encouraged in some treatment processes, denitrification often impairs the settling of the floc causing poor quality effluent in many suspended aeration plants. Overflow from the activated sludge mixing chamber is sent to a secondary clarifier where the suspended biological floc settles out while the treated water moves into tertiary treatment or disinfection. Settled floc is returned to the mixing basin to continue growing in primary effluent. Like most ecosystems, population changes among activated sludge biota can reduce treatment efficiency. Nocardia, a floating brown foam sometimes misidentified as *sewage fungus*, is the best known of many different fungi and protists that can overpopulate the floc and cause process upsets. Elevated concentrations of toxic wastes including pesticides, industrial metal plating waste, or extreme pH, can kill the biota of an activated sludge reactor ecosystem.

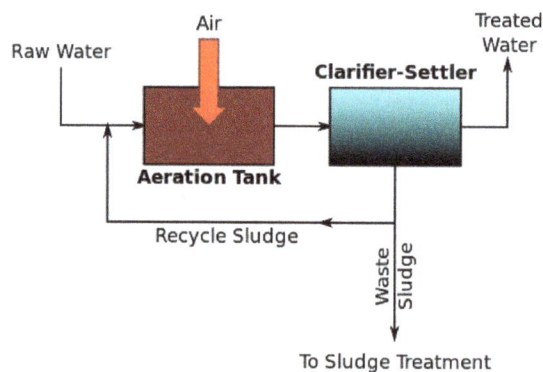

A generalized schematic of an activated sludge process.

Package Plants and Sequencing Batch Reactors

One type of system that combines secondary treatment and settlement is the cyclic activated sludge (CASSBR), or sequencing batch reactor (SBR). Typically, activated sludge is mixed with raw incoming sewage, and then mixed and aerated. The settled sludge is run off and re-aerated before a proportion is returned to the headworks.

The disadvantage of the CASSBR process is that it requires a precise control of timing, mixing and aeration. This precision is typically achieved with computer controls linked to sensors. Such a complex, fragile system is unsuited to places where controls may be unreliable, poorly maintained, or where the power supply may be intermittent. Extended aeration package plants use separate basins for aeration and settling, and are somewhat larger than SBR plants with reduced timing sensitivity.

Package plants may be referred to as *high charged* or *low charged*. This refers to the way the biological load is processed. In high charged systems, the biological stage is presented with a high organic load and the combined floc and organic material is then oxygenated for a few hours before being charged again with a new load. In the low charged system the biological stage contains a low organic load and is combined with flocculate for longer times.

Membrane Bioreactors

Membrane bioreactors (MBR) are activated sludge systems using a membrane liquid-solid phase separation process. The membrane component uses low pressure microfiltration or ultrafiltration membranes and eliminates the need for a secondary clarifier or filtration. The membranes are typically immersed in the aeration tank; however, some applications utilize a separate membrane tank. One of the key benefits of an MBR system is that it effectively overcomes the limitations associated with poor settling of sludge in conventional activated sludge (CAS) processes. The technology permits bioreactor operation with considerably higher mixed liquor suspended solids (MLSS) concentration than CAS systems, which are limited by sludge settling. The process is typically operated at MLSS in the range of 8,000–12,000 mg/L, while CAS are operated in the range of 2,000–3,000 mg/L. The elevated biomass concentration in the MBR process allows for very effective removal of both soluble and particulate biodegradable materials at higher loading rates. Thus increased sludge retention times, usually exceeding 15 days, ensure complete nitrification even in extremely cold weather.

The cost of building and operating an MBR is often higher than conventional methods of sewage treatment. Membrane filters can be blinded with grease or abraded by suspended grit and lack a clarifier's flexibility to pass peak flows. The technology has become increasingly popular for reliably pretreated waste streams and has gained wider acceptance where infiltration and inflow have been controlled, however, and the life-cycle costs have been steadily decreasing. The small footprint of MBR systems, and the high quality effluent produced, make them particularly useful for water reuse applications.

Aerobic Granulation

Aerobic granular sludge can be formed by applying specific process conditions that favour slow growing organisms such as PAOs (polyphosphate accumulating organisms) and GAOs (glycogen accumulating organisms). Another key part of granulation is selective wasting whereby slow settling floc-like sludge is discharged as waste sludge and faster settling biomass is retained. This process has been commercialized as Nereda process.

Surface-aerated Lagoons or Ponds

Aerated lagoons are a low technology suspended-growth method of secondary treatment using motor-driven aerators floating on the water surface to increase atmospheric oxygen transfer to the lagoon and to mix the lagoon contents. The floating surface aerators are typically rated to deliver the amount of air equivalent to 1.8 to 2.7 kg O_2/kW·h. Aerated lagoons provide less effective mixing than conventional activated sludge systems and do not achieve the same performance level. The basins may range in depth from 1.5 to 5.0 metres. Surface-aerated basins achieve 80 to 90 percent removal of BOD with retention times of 1 to 10 days. Many small municipal sewage systems in the United States (1 million gal./day or less) use aerated lagoons.

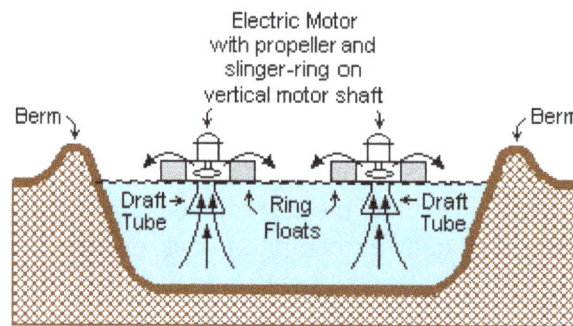

A TYPICAL SURFACE – AERATED BASIN

A typical surface-aerated basin (using motor-driven floating aerators)

Constructed Wetlands

Primary clarifier effluent was discharged directly to eutrophic natural wetlands for decades before environmental regulations discouraged the practice. Where adequate land is available, stabilization ponds with constructed wetland ecosystems can be built to perform secondary treatment separated from the natural wetlands receiving secondary treated sewage. Constructed wetlands resemble fixed-film systems more than suspended growth systems, because natural mixing is minimal. Constructed wetland design uses plug flow assumptions to compute the residence time required for treatment. Patterns of vegetation growth and solids deposition in wetland ecosystems, however, can create preferential flow pathways which may reduce average residence time. Measurement of wetland treatment efficiency is complicated because most traditional water quality measurements cannot differentiate between sewage pollutants and biological productivity of the wetland. Demonstration of treatment efficiency may require more expensive analyses.

Emerging Technologies

- Biological Aerated (or Anoxic) Filter (BAF) or Biofilters combine filtration with biological carbon reduction, nitrification or denitrification. BAF usually includes a reactor filled with a filter media. The media is either in suspension or supported by a gravel layer at the foot of the filter. The dual purpose of this media is to support highly active biomass that is attached to it and to filter suspended solids. Carbon reduction and ammonia conversion occurs in aerobic mode and sometime achieved in a single reactor while nitrate conversion occurs in anoxic mode. BAF is operated either in upflow or downflow configuration depending on design specified by manufacturer.

- Integrated Fixed-Film Activated Sludge.

- Moving Bed Biofilm Reactors typically requires smaller footprint than suspended-growth systems.

Tertiary Wastewaster Treatment

Tertiary treatment is the next wastewater treatment process after secondary treatment. This treatment is sometimes called as the final or advanced treatment and consists of removing the organic load left after secondary treatment for removal of nutrients from sewage and particularly to kill the pathogenic bacteria. The effluents from secondary sewage treatment plants contain both nitrogen (N) and phosphorus (P). N and P are ingredients in all fertilizers. When excess amounts of N and P are discharged, plant growth in the receiving waters may be accelerated which results in eutrophication in the water body receiving such waste. Algae growth may be stimulated causing blooms which are toxic to fish life as well as aesthetically unpleasing. Secondary treated effluent also contains suspended, dissolved, and colloidal constituents which may be required to be removed for stipulated reuse or disposal of the treated effluent.

The purpose of tertiary treatment is to provide a final treatment stage to raise the effluent quality before it is discharged to the receiving environment such as sea, river, lake, ground, etc., or to raise the treated water quality to such a level to make it suitable for intended reuse. This step removes different types of pollutants such as organic matter, SS, nutrients, pathogens, and heavy metals that secondary treatment is not able to remove. Wastewater effluent becomes even cleaner in this treatment process through the use of stronger and more advanced treatment systems. It includes sedimentation, coagulations, membrane processes, filtration, ion exchange, activated carbon adsorption, electrodialysis, nitrification and denitrification, etc. Tertiary treatment is costly as compared to primary and secondary treatment methods.

Need of Tertiary Treatment

- Tertiary treatment may be provided to the secondary effluent for one or more of the following contaminant further.

- To remove total suspended solids and organic matter those are present in effluents after secondary treatment.

- To remove specific organic and inorganic constituents from industrial effluent to make it suitable for reuse.

- To make treated wastewater suitable for land application purpose or directly discharge it into the water bodies like rivers, lakes, etc.

- To remove residual nutrients beyond what can be accomplished by earlier treatment methods.

- To remove pathogens from the secondary treated effluents.

- To reduce total dissolved solids (TDS) from the secondary treated effluent to meet reuse quality standards.

One or more of the unit operation/ process mentioned in Figure will be used for achieving this tertiary treatment.

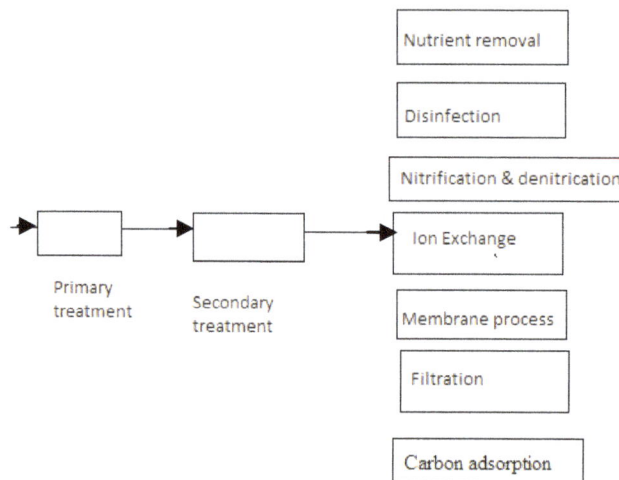

Process involved in tertiary wastewater treatment

Tertiary Treatments

In advanced wastewater treatment, treatment options or methods are dependent upon the characteristics of effluent to be obtained after secondary treatment to satisfy further use or disposal of treated wastewater.

Nitrogen Removal

Wastewater containing nutrients includes sewage, agriculture runoff and many of the industrial effluents. The nutrients of most concerned are N and P. The principal nitrogen compounds in domestic sewage are proteins, amines, amino acids, and urea. Ammonia nitrogen in sewage results from the bacterial decomposition of these organic constituents.

The nitrogen compounds results from the biological decomposition of proteins and from urea discharged in body waste. This nitrogen is in complex organic molecules and is referred simply as organic nitrogen. Organic nitrogen may be biologically converted to free ammonia (NH_3°) or to the ammonium ion (NH_4^+) by one of several different metabolic pathways. These two exists in equilibrium as

$$NH_4^+ \rightleftharpoons NH_3 + H^+$$

Ammonia nitrogen is the most reduced nitrogen compound found in wastewater, which can be biologically oxidized to nitrate if molecular oxygen is present (under aerobic condition). In wastewater, the predominant forms of nitrogen are organic nitrogen and ammonia. The nitrification may takes place in biological treatment units provided the treatment periods are long enough. Generally, for the HRT used in secondary treatment conversion of organic nitrogen to ammonia is significant and nitrification may not be significant. Because of oxygen demand exerted by ammonia (about 4.6 mg of O_2 per mg of NH^+-N oxidized) and due to other environmental factors, removal of ammonia may be required. The most common processes for removal of ammonia from wastewater are:

    i) Air stripping,

    ii) Biological nitrification and denitrification.

Air Stripping

It consists of converting ammonium to gaseous phase and then dispersing the liquid in air, thus allowing transfer of the ammonia from wastewater to the air. The gaseous phase NH_3^o and aqueous phase NH_4^+ exist together in equilibrium. The relative abundance of these phases depends upon both the pH and the temperature of the wastewater. The pH must be greater than 11 for complete conversion to NH_3. Since, this pH is greater than pH of normal wastewater, adjustment of pH is necessary prior to air stripping. Addition of lime is the most common means for raising the pH. Enough lime must be added to precipitate the alkalinity and to add the excess OH^- ions for pH adjustment.

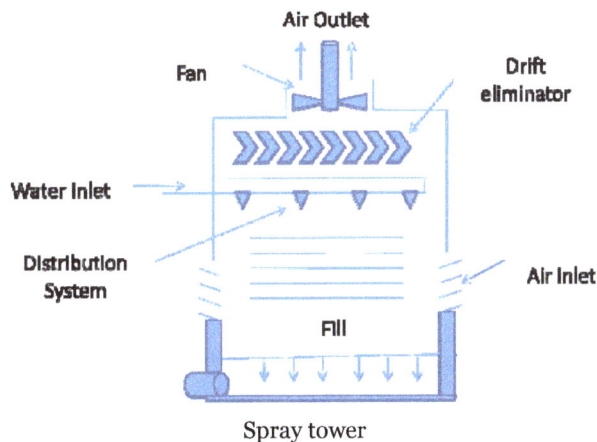

Spray tower

The most important and efficient reactor for air stripping is counter current spray tower. Larger quantity of air is required, and fan must be installed to draw air through tower. Packing is provided to minimize the film resistance to gas transfer by continuously forming, splashing and reforming drops.

The air to wastewater ratio ranging from 2000 to 6000 m³ of air/m³ of wastewater is used for design. Air requirement is more at lower temperature. Tower depths are generally less than 7.5 m, and hydraulic loading vary from 40 to 46 L/min.m² of tower. The limitation to this process occurs at temperature close to freezing temperature. Drastic reduction in efficiency is observed and preheating of gas is not possible because of high volume.

The other problems associated with this include noise and air pollution and scaling of the packing media. Noise pollution problem is caused by roar of the fan. The odor problem is due to dispersion of ammonia gas in atmosphere, due to which this technique may not be permitted at some location. This problem can be minimized by locating the facility away from the populated area. Precipitation of calcium carbonate scale on the packing media as a result of wastewater softening can be minimized by the use of smooth surface PVC pipe as packing material. The occasional cleaning of packing media is still required.

Biological Nitrification and Denitrification

Bacteria remove ammonia nitrogen from wastewater by a two step biological processes: nitrification followed by denitrification to covert it finally to gaseous nitrogen. In this gaseous form N2 is inert and does not react with the wastewater itself or with other constituents present in wastewater. Since, treated wastewater is likely to be saturated with molecular nitrogen; the produced N2 is simply released to the atmosphere. These two steps involved require different environmental conditions and hence generally they are carried out in separate reactors.

Nitrification

It has important role in nitrogen removal from wastewater during treatment. The biological conversion of ammonium to nitrate nitrogen is called Nitrification. It is autotrophic process i.e. energy for bacterial growth is derived by oxidation of nitrogen compounds such as ammonia. In this process, the cell yield per unit substrate removal is smaller than heterotrophs. Nitrification is a two-step process. In first step, bacteria known as *Nitrosomonas* can convert ammonia and ammonium to nitrite. These bacteria known as nitrifiers are strictly aerobes. This process is limited by the relatively slow growth rate of *Nitrosomonas*. Next, bacteria called *Nitrobacter* finish the conversion of nitrite to nitrate.

Nitrosomonas

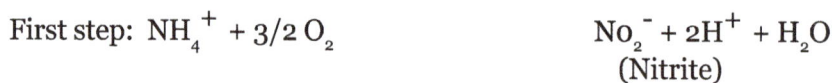

First step: $NH_4^+ + 3/2\,O_2 \longrightarrow NO_2^- + 2H^+ + H_2O$ (Nitrite)

Nitrobacter

Second step: $NO_2^- + 1/2\,O_2 \longrightarrow$ (Nitrate)

Overall reaction: $NH_4^+ + 2\,O_2 \longrightarrow NO_3^- + 2H^+ + H_2O$

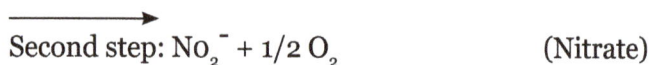

Nitrosomonas and *Nitrobacter* use the energy derived from the reactions for cell growth and maintenance. Some of ammonium ions are assimilated into cell tissues. Neglecting this ammonium ion used in cell synthesis the O_2 required to oxidize ammonia to nitrate is 4.57 mg O_2/mg ammonium nitrogen. If the ammonium used in cell, O2 required is considered it is 4.3 mg O_2/mg ammonium nitrogen and about 7.14 mg of alkalinity is needed to neutralize the H+ produced.

Nitrification may be used to prevent oxygen depletion from nitrogenous demand in the receiving water. Nitrification requires a long retention time, a low food to microorganism ratio (F/M), a high mean cell residence time (MCRT), and adequate alkalinity. Wastewater temperature and Ph affects the rate of nitrification.

Under favourable conditions, carbon oxidation and nitrification may occur in a single reactor called single stage under favourable conditions. In separate stage carbon oxidation and nitrification occur in different reactors. It can be accomplished in both suspended and attached growth process such as trickling filter, ASP, rotating disc biological contactor (RBC), SBR, etc. Also, nitrifying organisms are present in almost all aerobic process sludge but they are less in number. In nitrification, when ratio of BOD_5 to TKN is greater than 5, the process is called as combined carbon oxidation and nitrification, whereas, when ratio of BOD_5 to TKN is less than 3, the process is called as separate stage carbon oxidation and nitrification.

For single stage carbon oxidation and nitrification, plug flow, completely mixed and extended aeration ASP can be used. Oxidation ditch is also one of option for this process. The attached growth processes like TF and RBC are commonly used. Nitrification can be achieved by reducing applied loading rate, increasing the mean cell residence time (θ_c) and air supply.

Nitrifying bacteria are sensitive organisms. A variety of organic and inorganic agents can inhibit the growth and action of these organisms. High concentration of ammonia and nitrous acid can be inhibitory. The effect of pH is also significant with optimal range of 7.5 to 8.6. The system acclimatize to lower pH can also work successfully. The temperature also has considerable impact on growth of the nitrifying bacteria. Dissolved oxygen concentration above 1 mg/L is essential for nitrification. Below this DO, oxygen becomes the limiting nutrients and nitrification slows down or ceases.

Denitrification

In some applications, such as discharge of effluent into enclosed water bodies or recycle to water supplies, nitrification may not be sufficient. When nitrogen removal is required, one of the available methods is to follow biological nitrification with denitrification. Denitrification is accomplished under anaerobic or near anaerobic conditions by facultative heterotrophic bacteria commonly found in wastewater. Nitrates are removed by two mechanisms: (1) conversion of NO_3 to N_2 gas by bacterial metabolism and (2) conversion of NO_3 to nitrogen contained in cell mass which may be removed by settling. Denitrification occurs when oxygen levels are depleted and nitrate becomes the primary electron acceptor source for microorganisms.

$$\text{Nitrate}, NO_3^- \rightarrow \text{Nitrate}, NO_2^- \rightarrow \text{Nitrate oxide}, NO \rightarrow \text{Nitrous oxide}, N_2O \rightarrow \text{Nitrogen}, N_2$$

Denitrifying bacteria are facultative organisms, they can use either dissolved oxygen or nitrate as an oxygen source for metabolism and oxidation of organic matter. This is carried out by hetetrophic bacteria such as *pseudomonas, spirillum, lactobacillus, bacillus, microaoccus*, etc. For reduction to occur, the DO level must be near to zero, and carbon supply must be available to the bacteria. Because of low carbon content is required for the previous nitrification step, carbon must be added before denitrification can proceed. A small amount of primary effluent, bypassed around secondary and nitrification reactor can be used to supply the carbon. However, the unnitrified compounds in this water will be unaffected by the denitrification process and will appear in effluent.

When complete nitrogen removal is required, an external source of carbon containing no nitrogen will be required. The most commonly used external source of nitrogen is methanol. When methanol is added the reaction is

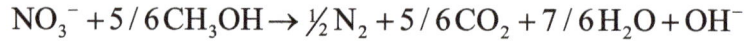

$$NO_3^- + 5/6\,CH_3OH \rightarrow \tfrac{1}{2}N_2 + 5/6\,CO_2 + 7/6\,H_2O + OH^-$$

For treatment plant above 3 mg/L of methanol is required for each milligram per litre of nitrate, making this process an expensive. Alkalinity is generated in this process. Denitrification can be carried out as attached growth (anaerobic filter) and suspended growth process (expanded bed or UASB reactor).

Phosphorus Removal

Normally secondary treatment can only remove 1-2 mg/l of phosphorus, so a large excess of phosphorus is discharged in the final effluent, causing eutrophication of lakes and natural water bodies. Generally it appears as orthophosphate, polyphosphate and organically bound phosphorus. Phosphorus is required for microbes for cell synthesis and energy transport, maintenance as well as it is stored for subsequent use by microbes. During secondary treatment process about 10 to 30 % of influent phosphorus is removed (Metcalf & Eddy, 2003). Phosphate removal is currently achieved largely by chemical precipitation, which is expensive and causes an increase of sludge volume by up to 40%. An alternative is the biological phosphate removal (BPR), which is accomplished by sequencing and producing the appropriate environmental condition in the reactors.

Acinetobacter organisms are helpful for removal of phosphorus from effluent. Under anaerobic conditions, phosphorus accumulating organisms (PAO) assimilate fermentation products (i.e. volatile fatty acids) into storage products within the cells with the concomitant release of phosphorus from stored polyphosphates (Gray, 2005). Release of phosphorus occurs under anoxic condition. The BPR requires both aerobic and anaerobic zones in reactors for efficient treatment. Generally, lime precipitation is most commonly used for phosphorus removal. Phosphorus is removed in the waste sludge from the system.

Treatment technologies presently available for phosphorus removal include (de-Bashan and Bashan, 2004):

A) Physical:

    a) Filtration for particulate phosphorus

    b) Membrane technologies

B) Chemical:

    a) Precipitation

    b) Other (mainly physical-chemical adsorption)

C) Biological:

    a) Assimilation

    b) Enhanced biological phosphorus removal (EBPR)

Ion Exchange

Ion Exchange can be used in wastewater treatment plants to swap one ion for another for the purpose of demineralization. The widest application of this process is in domestic water softening, where sodium ions are removed on cation exchange resin and chlorides are removed on anion exchange resin. Ion exchange is a unit process in which ions are removed out of an aqueous solution and is replaced by another ionic species. The basic principle behind ion exchange is that a weakly bound ion can preferably be displaced by a stronger binding ion. This effect is called the principle of selectivity. A more selective ion binds more strongly than a less selective ion. The effect of selectivity can be used to remove distinct ions from water and to replace them with others. It can be operated in a batch or continuous mode and has been used for removal of nitrogen, heavy metals and TDS in wastewater applications. It has also been used selectively to remove specific impurities and to recover valuable trace metals like chromium, nickel, copper, lead and cadmium from industrial waste discharges.

A number of naturally occurring minerals have ion exchange properties. Among them the notable ones are aluminium silicate minerals, which are called zeolites. Synthetic zeolites have been prepared using solutions of sodium silicate and sodium aluminate. Alternatively synthetic ion-exchange resins composed of organic polymer with attached functional groups such as $-SO_3^-$ H^+ (strongly acidic cation exchange resins), or $- COO^- - H^+$ (weakly acidic cation exchange resins or $-N^+(CH_3)_3OH^-$ (strongly basic anion exchange resins) can be used. Synthetic and industrially produced ion exchange resins consist of small, porous beads that are insoluble in water and organic solvents. The most widely used base-materials are polystyrene and polyacrylate.

Membrane Process

Membrane technology can be used to treat a variety of wastes, including sanitary landfill leachate containing both organic and inorganic chemical species, water-soluble oil wastes used in metal fabricating and manufacturing industries, solvent-water mixtures, and oil-water mixtures generated during washing operations at metal fabricating facilities. Depending upon the material used for membrane, nature of driving force and separation mechanism, the membrane processes can be classified into sub-processes such as electrodialysis (ED) or electrodialysis reversal (EDR), microfiltration (MF), ultrafiltration (UF), nanofiltration (NF), reverse osmosis (RO).

MF, UF, NF and RO use pressure to transport water across the membrane. MF removes particulate matter while RO removes many solutes as water permeate through the membrane. MF membranes have the largest pore size and typically reject large particles and various microorganisms. UF membranes have smaller pores than MF membranes and, therefore, in addition to large particles and microorganisms, they can reject bacteria and soluble macromolecules such as proteins. RO membranes are effectively non-porous and, therefore, exclude particles and even many low molar mass species such as salt ions, organics, etc. NF membranes are relatively new and are sometimes called loose RO membranes. They are porous membranes, but since the pores are of the order of ten angstroms or less, they exhibit performance between that of RO and UF membranes. Different membrane processes are being used in water treatment since 1960.

The driving forces for the use of membrane technology are increased regulatory pressure to provide better quality water, increased demand of water requiring exploitation of low quality water resources as source and development and commercialization of membrane processes. Application of membrane processes for wastewater treatment is increasing worldwide day by day with reduction in cost of process and increased water charges. Membrane processes are used for about 70 % of the total installed capacity of desalination worldwide and percentage is increasing (for drinking water). In Japan, more than 30 MLD wastewater is used as toilet flushing water in building. It has application in many industrial wastewater treatment plants to produce reusable quality water from the effluents.

Types of Membrane Operation

The membrane is defined as the thin film separating two phases and acting as a selective barrier to the transport matter. Chemical potential difference exists between the two phases. *Retentate* contains non permeating species. *Permeate* forms the produced water from the membrane filtration. Membrane operation is recommended nomenclature than membrane process. The relevant main membrane operations used in water treatment are summarized in Table.

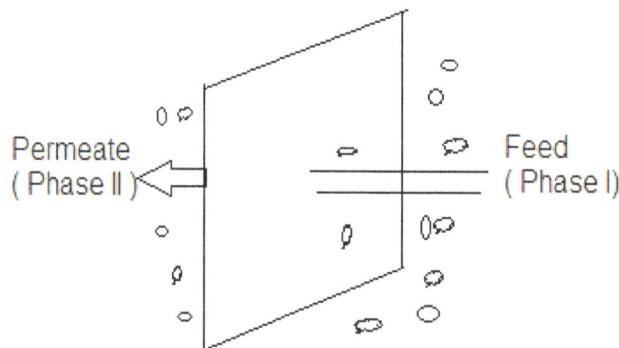

Mechanism of membrane operation

Table: Technically relevant main membrane operations in water treatment

Membrane	Driving force	Mechanism of separation	Membrane structure	Phases in contact
Microfiltration (MF)	pressure	Sieve	Macropores	Liquid-liquid
Ultrafiltration (UF)	pressure	Sieve	Mesopores	Liquid-liquid
Nanofiltration (NF)	pressure	Sieve + solution + diffusion + exclusion	Micropores	Liquid-liquid
Reverse Osmosis	pressure	Solution/diffusion + exclusion	Dense	Liquid-liquid

Advantages of membrane operation:

- Separation takes place at ambient temperature without phase change.
- Separation takes place without accumulation of product inside membrane (unlike ion exchange resins, which needs replacement/regulations).
- No need of chemical additives for separations.

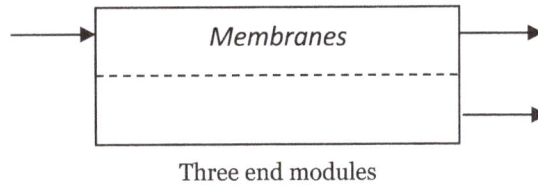

Three end modules

Membrane Operations

The transmembrane flux for each element is given by

Flux = Force * concentration * mobility

In most cases, concentration will vary with distance from the membrane surface along the boundary layer. Hence above equation is local equation where local forces are used to estimate gradient of chemical potential $d\mu/dx$ of every component that can be transported.

The variation of chemical potential of component 'i' can be expressed as a sum of the terms

$$d\mu_i = RTd\,lna_i + V_i\,dP + Z_i\,Fd\Psi$$

where, activity a_i (product of concentration by activity coefficient) is not under the control of operator. The pressure P and electric potential Ψ can be varied in order to improve the separation between the mobile components. The applied pressure of every component is proportional to its molar volume V_i. Electric field acts on every ionic species according to its valency Z_i and does not affect non-ionic species.

Reverse Osmosis

Reverse osmosis (RO) is a most commonly used membrane filtration method that removes many types of large molecules and ions from effluents by applying pressure to the effluents when it is on one side of a selective membrane. RO is used to remove specific dissolved organic constituents from wastewater remaining after advanced treatment with depth filtration or MF. RO system can operate at 90 % efficiency or better with respect to TDS. In addition, it also removes residual organic molecules, turbidity, bacteria and viruses.

The solvent of the solution is transferred through a dense membrane tailored to retain salts and low molecules weight solutes. When concentrated saline solution is separated from pure water by such a membrane, difference in chemical potential tends to promote the diffusion of water from the diluted compartment to the concentrated compartment in order to equalize concentration. At equilibrium, the difference in the levels between two compartments corresponds to the osmotic pressure of the saline solution. The demineralization of water can be accomplished using thin, microporous membranes. There are two basic modes of operation in use. One system uses pressure to drive water through the membrane against the force of osmotic pressure and is called RO. The pressure applied is several orders of magnitude in excess of natural osmotic pressure. The membrane commonly used in reverse osmosis is composed of cellulose acetate and is about 100 µm thick. The film contains microscopic openings that allow water molecules to pass through but retain dissolved solids by either molecular sieving (or by some other mechanism which is not yet completely understood). The process results in a concentrated solution of the ions on the pressure side of the membrane and product water relatively free of ions on the other side of the membrane.

For the producing pure water from the saline solution, osmotic pressure of the solution must be exceeded in the brine. For economically viable flows, at least twice the osmotic pressure must be exerted e.g. for seawater pressure of 5 -8 MPa is used in practice.

Nanofiltration (NF)

Nanofiltration (NF) is a rapidly advancing membrane separation technique for water and wastewater treatment as well as concentration/separation of antibiotics and pharmaceuticals due to its unique charge-based repulsion property and high rate of permeation. This process is also called as low pressure RO or membrane softening. It lies between RO and ultrafiltration in terms of selectivity of membranes, which is designed for removal of multivalent ions (Ca, Mg, etc.) in softening operation. Monovalent ions are poorly rejected. Hence, osmotic back pressure is much lower than that in RO. The operating pressure used in NF is typically 0.5 to 1.5 MPa.

It is a pressure driven process wherein the pore size of the membrane is in the range of 0.5-1 nm. Due to the lower operating pressure and higher flow rates, nanofiltration is inexpensive when compared to reverse osmosis. NF membranes allow partial permeation of monovalent salts such as sodium chloride while rejecting bivalent salts and hardness to a greater extent from aqueous solutions. NF can lower TDS and hardness, reduce color and odor, and remove heavy metal ions from groundwater. Other possible applications include treatment of effluents from textile dyeing, bulk drug and chemical process industries.

NF process is useful for raw water containing TDS in the range 700-1200 ppm and has exhibited two advantages over RO. NF allows sufficient permeation of essential salts while keeping the total dissolved solids (TDS) in accordance to BIS drinking water standards, whereas RO removes almost all the minerals essential for the human body. NF can be operated at lower pressure while RO requires higher operating pressure and therefore higher running cost.

Ultrafiltration (UF)

It is a clarification and disinfection membrane operation. UF membranes are porous and allow only coarser solutes (macromolecules) to be rejected. All types of microorganisms as viruses and bacteria and all types of particles can be removed by this process. Since, the low molecular solutes are not retained by UF, osmotic back pressure can be neglected and operating pressure is kept low as 50- 500 KPa.

Microfiltration (MF)

Microfiltration (MF) membranes are having 0.1 µm or more pore size. It is generally used for particulate matter removal. The pressure used in this process is similar to that of UF.

Classification of Membranes

According to separation mechanism: Three mechanisms exist for separation of ions and other particulate matter from the membranes, and accordingly the membranes are classified as below:

- Sieve effect: In this mechanism, separation is based on difference in pore size e.g. MF, UF.

- Solution-diffusion mechanism: In this mechanism, separation is based on difference in the solubility and diffusivity of materials in the membrane e.g. RO.

- Electrochemical effect: In this mechanism, separation is based on difference in the charges of the species to be separated e.g. ED (electrodialysis).

The classification based on separation mechanism leads to three main classes of membranes.

i) Porous membranes

Fixed pores are present in these membranes. These pores are sub-divided into three types' viz.

- Macropores: Theses are larger than 50 nm

- Mesopores- These are in the range of 20 to 50 nm

- Micropores- These are in the size less than 2 nm.

MF and UF are porous membranes while, NF could be classified in an intermediate class between porous and non porous membranes. Since, solution diffusion and even electrochemical effect have to introduce in equilibrium of mass transfer.

ii) Non Porous membranes

These are dense media membranes. The diffusion of species takes place in the free volume which is present between the macromolecular chains of the membrane material. RO is non porous membrane.

iii) Ion exchange membranes

These are specific types of non porous membranes. They consist of highly swollen gels carrying fixed positive or negative charges. A membrane with fixed positive charges (e.g. - NR_3^+) is called as anion exchange membranes, whereas cation exchange membranes have fixed negative charges (e.g. $- SO^{3-}$).

Classification based on Morphology

i) Asymmetric membranes

These are anisotropic membranes prepared from the same material.

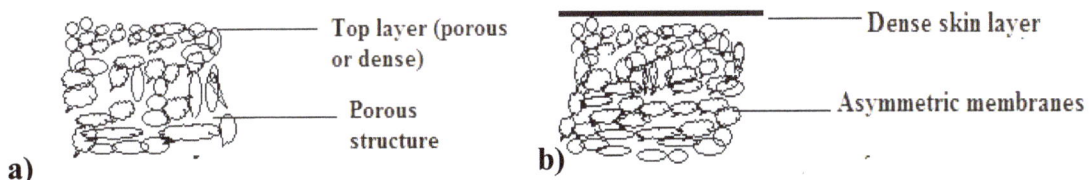

a) Asymmetric membrane and b) Composite membrane.

ii) Composite membranes

These are anisotropic membranes where top layer and sub-layer originate from different materials. Each layer can be optimized independently. Generally porous layer is asymmetric membranes.

The anisotropic membrane consists of very thin top layer called skin supported by a thicker and more porous supporting sub-layer. Skin has main function of membrane. Overall flux and selectivity depends upon the structure of skin. Skin has thickness in the range of 0.1 to 0.5 μm, which is about 1% of the thickness of porous sub-layer. Supporting layer offers negligible resistance to mass transfer and imparts mechanical strength to the membrane forming integral part of the membrane.

Classification based on Geometry

According to the geometry, the membranes can be classified as flat sheet membrane and cylindrical membrane. The cylindrical membrane can be further classified as (a) Tubular membranes having internal diameter >3 mm, and (b) Hollow fiber membranes having internal diameter < 3 mm. They are available in market with outer diameter ranging from 80 to 500 μm. They are used in RO with larger diameter. Also hallow fiber membranes are used in MF and UF in which they are also called as 'capillary membranes'.

Classification based on Chemical Nature

On the basis of chemical nature of the material used for membrane these are classified as: organic (polymers) membrane and inorganic membranes made from metals, ceramics, glasses, etc.

a) Polymer (organic) membranes

Cellulose and its derivatives are more widely used. These hydrophilic polymers are low cost and they are having low tendencies for adsorption. Polyamides, a hydrophilic polymers, are second type of polymer (aromatic polyamides) used for making membrane after cellulose diacetate. It has better thermal, chemical and hydraulic stability than cellulose esters. Amide group (-CO-NH-) cannot tolerate exposure to trace of Cl2, hence they are not suitable for chlorinated water. Poly acrylonitrile (PAN) are used in UF but not in RO.

Polysulphone (PSF) and Poly ethersulphone (PES) - These two are not hydrophilic and hence they have high adsorption tendency. They have very good thermal, chemical and mechanical stability. Polytetra fluoro ethylene (PTFE), Polyvinylidene fluoride (PVDF), Polyethylene (PE), Polycarbonate (PC) or isotactic Polypropylene (PP) are hydrophilic polymers that are used for membranes.

b) Inorganic membranes

These membranes have superior thermal, chemical and mechanical stability relative to polymer materials. The disadvantage of this membrane is that they are brittle and more expensive. Ceramic membranes are oxides, nitrides or carbides of metals such as Al, Zr, Ti, etc.

References

- Tchobanoglous, G., Burton, F.L., and Stensel, H.D. (2003). Wastewater Engineering (Treatment Disposal Reuse) / Metcalf & Eddy, Inc (4th ed.). McGraw-Hill Book Company. ISBN 0-07-041878-0

- "Ultraviolet light disinfection in the use of individual water purification devices" (PDF). U.S. Army Public Health Command. Retrieved 2014-01-08

- Omelia, C (1998). "Coagulation and sedimentation in lakes, reservoirs and water treatment plants". Water Science and Technology. 37 (2): 129. doi:10.1016/S0273-1223(98)00018-3

- Messina, Gabriele (October 2015). "A new UV-LED device for automatic disinfection of stethoscope membranes" (PDF). American Journal of Infection Control. Elsevier. Retrieved 2016-08-15

- Franson, Mary Ann Standard Methods for the Examination of Water and Wastewater 14th edition (1975) APHA, AWWA & WPCF ISBN 0-87553-078-8 p.131

- Ware, M. W.; et al. "Inactivation of Giardia muris by low pressure ultraviolet light" (PDF). United States Environmental Protection Agency. Archived from the original (PDF) on 27 February 2008. Retrieved 2008-12-28

- Meulemans, C. C. E. (1987-09-01). "The Basic Principles of UV–Disinfection of Water". Ozone: Science & Engineering. 9 (4): 299–313. doi:10.1080/01919518708552146. ISSN 0191-9512

- Ware, M. W.; et al. "Inactivation of Giardia muris by low pressure ultraviolet light" (PDF). United States Environmental Protection Agency. Archived from the original (PDF) on 27 February 2008. Retrieved 2008-12-28

- Downes, Arthur; Blunt, Thomas P. (19 December 1878). "On the Influence of Light upon Protoplasm" (PDF). Proceedings of the Royal Society of London. 28 (190–195): 199–212. doi:10.1098/rspl.1878.0109. Retrieved 4 May 2015

- C., Reed, Sherwood (1988). Natural systems for waste management and treatment. Middlebrooks, E. Joe., Crites, Ronald W. New York: McGraw-Hill. pp. 268–290. ISBN 0070515212. OCLC 16087827

Chapter 4

Wastewater Manangement Systems

Science and technology has undergone rapid development in the past decade, which has resulted in the emergence of many innovative tools and techniques for wastewater management. These include effective wastewater management systems such as onsite sewage facility, aerated lagoon, decentralized wastewater system, grinder pump and grease trap, which have been extensively detailed in this chapter.

Onsite Sewage Facility

An On-Site Sewage Facility (OSSF) is a wastewater system that treats and disposes of sewage produced on the site location and has a daily usage of less than 5000 gallons. An OSSF is comprised of two components: the treatment and the disposal. These two components may be set up in a variety of different scenarios including, but not limited to, a septic tank with lateral lines for subsurface treatment, an aerobic treatment unit with surface spray application of treated and disinfected wastewater, or several other methods of waste water disposal. A properly functioning OSSF should drain the plumbing (your toilets flush correctly), have no pooling or ponding, have no odor, and should provide nearly 100% wastewater treatment.

The most common and traditional septic systems consist of a septic tank that gravity flows to a soil adsorption field for final treatment and dispersal. The septic tank allows particulate matter to settle to the bottom of the tank so that large solids do not plug the drain field. An effluent screen placed in the outlet of the septic tank is used to filter suspended solids out of the effluent. Final treatment and dispersal of the wastewater takes place in the soil adsorption filed.

A non-traditional system performs the same basic actions as the conventional septic systems. Differences arise when location, space, laws and regulations, soil type, and/or quantity of wastewater being treated become a limiting factor. A non-traditional system in this case refers to any OWTS that uses pumps or advanced treatment. These systems use technologies that require greater frequency of operation and maintenance.

An OWTS can be divided 4 main components:

1. Wastewater Source

2. Collection and Storage

3. Pretreatment Components

4. Final Treatment and Dispersal Components

Wastewater Source (Homeowners)

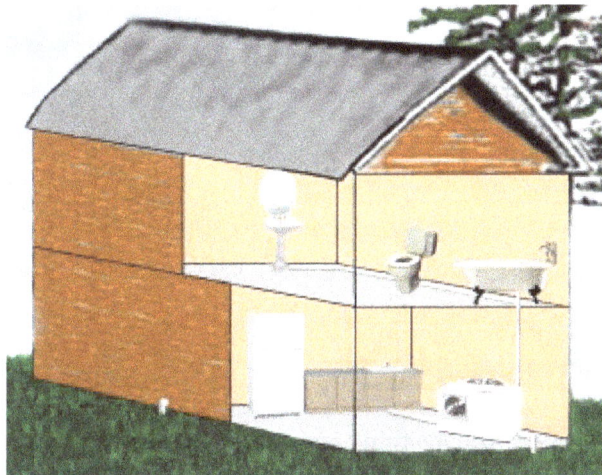

Wastewater is composed of 99.9% water. The constituents and strength of wastewater can vary depending on the source. Domestic wastewater is the water that is discharged from plumbing fixtures, appliances, toilets, bath, laundry, and dishwashers in a residence.

Commercial wastewater comes from businesses, restaurants and manufacturing plants. Septic systems owners should be aware that anything they dump down their sinks or drains will end up passing through their septic system and end up in their ground water and surface water supplies.

Homeowners also should be aware of the operation and maintenance requirements of their system. Proper operation and maintenance can protect public and environmental health, increase their systems longevity, as well as protect property use and value. The Homeowners Guide to Evaluating A Service Contract is a valuable tool for septic system owners who choose to have a professional take care of their system.

The constituents that are present in wastewater include:

- Organics

- Inorganics

- Solids

- Pathogenic Organisms

- Nutrients

- Metals

- Persistent Organic Compounds

- Fats, Oils, and Grease (FOG)

- Endocrine Disputers

Collection and Storage

Collection components of residential systems are generally limited to a solid, rigid pipe collecting wastewater from plumbing fixtures and appliances. This pipe, laid on a 1-2 percent downward slope (¼ in per ft), exits the structure, and extends to the pretreatment component.

A clean-out should be located in the pipe before the first pretreatment component. Depending on sampling needs and requirements, adequate sampling basins should be located between components.

Some sites may have elaborate alternative collection systems. These systems may have pump tanks collecting the waste and subsequent transmission lines for transporting it to the pretreatment components. Holding tanks can be considered storage devices, because they essentially store wastewater until it is collected and transported to a different site for treatment and dispersal.

Components:

- Holding Tank

- Alternative Collection System

- Pump Tank

- Grease Trap

- Incinerating Toilet

- Composting Toilet

- Graywater System

Pretreatment Components

Pretreatment components remove many of the contaminants from the wastewater to prepare the effluent for final treatment and dispersal into the environment. The level of treatment is selected to match the receiving environment and the intended use.

The quantity of contaminants is reduced to a level the soil can accept and treat. Many options exist for treatment prior to release into the receiving environment. Wastewater pretreatment components include septic tanks, trash tanks, and processing tanks, while aerobic treatment units, media filters, and constructed wetlands are considered advanced pretreatment components.

Components:

- Septic Tank
- Trash Tank
- Processing Tank
- Effluent Screen
- Recirculation Tank

Final Treatment and Dispersal

Final treatment and dispersal components provide the final removal of contaminants and distribute the effluent for dispersal back into the environment. Several options are available for distributing wastewater in soil. Gravity flow systems are the most widely used dispersal systems. These systems will continue to be used in areas where the soil separation distances can be met, primarily because they are the least expensive alternative and require the least amount of operation and maintenance.

Pressurized distribution methods overcome a variety of site limitations. Low pressure, subsurface drip, and spray distribution systems are designed to function in difficult areas. These systems are pressurized, which assists in providing even distribution of wastewater. These technologies also facilitate reuse of wastewater in the landscape. These advantages, however, increase the operation and maintenance requirements.

Methods

- Soil Adsorption Field
- Conventional Drainfield System
- Gravel-less Pipe
- Leaching Chamber
- Mound System
- Low-Pressure Drainfield (LPD)
- Low-Pressure Pipe (LPP)
- Shallow Narrow Drainfield
- Spray Distribution
- Drip Distribution
- Evapotranspiration Bed (ET)
- Media Filter as Drainfield Option
- Bottomless Sand Filter
- Bottomless Peat Filter

Pretreatment has two Main Categories

1. Advanced Pretreatment
2. Disinfection

Advanced Pretreatment Components

Advanced wastewater treatment increases the percentage of contaminants, particularly nitrogen and fecal coliform, removed in wastewater. Advanced pretreatment components typically follow primary treatment from septic tanks and decrease the constituents of concern before they reach the final treatment and dispersal component. Advanced pretreatment components are used when a site has a high risk to public or environmental health and primary treatment is not protective enough.

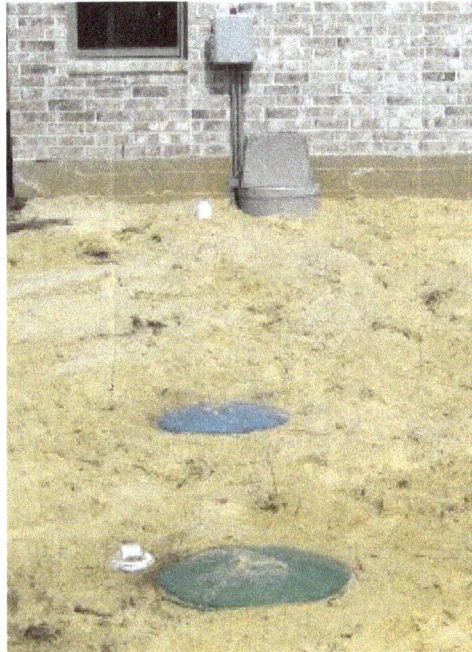

Advanced Pretreatment Components Include

- Aerobic treatment units (ATUs)

- Constructed wetlands

- Lagoons

- Media filters:

  ◦ *Trickling Filter*

  ◦ *Sand/Gravel Filter*

  ◦ *Foam Filter*

  ◦ *Peat Filter*

  ◦ *Textile Filter*

  ◦ *Upflow Filter*

Disinfection

Onsite wastewater treatment systems distributing wastewater on the ground surface are required to include a disinfection component as part of the advanced pretreatment process. Additionally, some subsurface drip systems applying wastewater into shallow soils require disinfection prior to dispersal. Disinfection is the destruction or inactivation of disease-causing organisms. The disinfection component reduces the concentration of the pathogenic constituents to an acceptable level. This usually relates to a health standard or a maximum required number of organisms for infection.

Wastewater can be disinfected with many methods. Chlorine, ultraviolet light, and ozone will be discussed to provide a greater understanding of how they operate. For onsite wastewater treatment systems, the most common form of disinfection is chlorination.

Methods

- Chlorination
    - *Tablet*
    - *Liquid*
- Ultraviolet Light
- Ozone
- Dechlorination

Improving Treatment through Performance Requirements

Most onsite wastewater treatment systems are of the conventional type, consisting of a septic tank and a subsurface wastewater infiltration system (SWIS). Site limitations and more stringent performance requirements have led to significant improvements in the design of wastewater treatment systems and how they are managed. Over the past 20 years the onsite wastewater treatment system (OWTS) industry has developed many new treatment technologies that can achieve high performance levels on sites with size, soil, ground water, and landscape limitations that might preclude installing conventional systems. New technologies and improvements to existing technologies are based on defining the performance requirements of the system, characterizing wastewater flow and pollutant loads, evaluating site conditions, defining performance and design boundaries, and selecting a system design that addresses these factors.

Performance requirements can be expressed as numeric criteria (e.g., pollutant concentration or mass loading limits) or narrative criteria (e.g., no odors or visible sheen) and are based on the assimilative capacity of regional ground water or surface waters, water quality objectives, and public health goals. Wastewater flow and pollutant content help define system design and size and can be estimated by comparing the size and type of facility with measured effluent outputs from similar, existing facilities. Site evaluations integrate detailed analyses of regional hydrology, geology, and water resources with site specific characterization of soils, slopes, structures, property lines, and other site features to further define system design requirements and determine the physical placement of system components.

Most of the alternative treatment technologies applied today treat wastes after they exit the septic tank; the tank retains settleable solids, grease, and oils and provides an environment for partial digestion of settled organic wastes. Post-tank treatment can include aerobic (with oxygen) or anaerobic (with no or low oxygen) biological treatment in suspended or fixed-film reactors, physical/chemical treatment, soil infiltration, fixed-media filtration, and/or disinfection. The application and sizing of treatment units based on these technologies are defined by performance requirements, wastewater characteristics, and site conditions.

Usage

In The United States

In the United States, on site sewage facilities collect, treat, and release about 4 billion US gallons (15,000,000 m³) of treated effluent per day from an estimated 26 million homes, businesses, and recreational facilities nationwide (U.S. Census Bureau, 1997). Recognition of the impacts of onsite systems on ground water and surface water quality (e.g., nitrate and bacteria contamination, nutrient inputs to surface waters) has increased interest in optimizing the systems' performance. Public health and environmental protection officials now acknowledge that onsite systems are not just temporary installations that will be replaced eventually by centralized sewage treatment services, but permanent approaches to treating wastewater for release and reuse in the environment. Onsite systems are recognized as viable, low-cost, long-term, decentralized approaches to wastewater treatment if they are planned, designed, installed, operated, and maintained properly (USEPA, 1997). In addition to existing state and local oversight, decentralized wastewater treatment systems that serve more than 20 people might become subject to regulation under the USEPA's Underground Injection Control Program, although EPA has proposed not to include them.

Although some onsite wastewater management programs have functioned successfully in the past, problems persist. Most current onsite regulatory programs focus on permitting and installation.

Few programs address onsite system operation and maintenance, resulting in failures that lead to unnecessary costs and risks to public health and water resources. Moreover, the lack of coordination among agencies that oversee land use planning, zoning, development, water resource protection, public health initiatives, and onsite systems causes problems that could be prevented through a more cooperative approach. Effective management of onsite systems requires rigorous planning, design, installation, operation, maintenance, monitoring, and controls.

Public Health and Water Resource Impacts

State and tribal agencies report that onsite septic systems currently constitute the third most common source of groundwater pollution and that these systems have failed because of inappropriate siting or design or inadequate long-term maintenance (USEPA, 1996a). In the 1996 Clean Water Needs Survey (USEPA, 1996b), states and tribes also identified more than 500 communities as having failed septic systems that have caused public health problems. The discharge of partially treated sewage from malfunctioning onsite systems was identified as a principal or contributing source of degradation in 32 percent of all harvest-limited shellfish growing areas. Onsite wastewater treatment systems have also contributed to an overabundance of nutrients in ponds, lakes, and coastal estuaries, leading to the excessive growth of algae and other nuisance aquatic plants (USEPA, 1996b). In addition, onsite systems contribute to contamination of drinking water sources. USEPA estimates that 168,000 viral illnesses and 34,000 bacterial illnesses occur each year as a result of consumption of drinking water from systems that rely on improperly treated ground water. Malfunctioning septic systems have been identified as one potential source of ground water contamination (USEPA, 2000).

Aerated Lagoon

In this section we will discuss about:- 1. Types of Aerated Lagoons 2. Design of the Aerated Lagoons 3. Advantages.

Types of Aerated Lagoons

Aerated lagoons are deep waste stabilization ponds in which sewage is aerated by mechanical aerators to stabilize the organic matter present in the sewage, rather than relying only on photosynthetic oxygen produced by algae. Thus aerated lagoons represent a system of sewage treatment that is intermediate between oxidation ponds and activated sludge systems.

Depending on how the microbial mass of solids is handled in the aerated lagoons the same are classified as:

  (i) Facultative aerated lagoons and

  (ii) Aerobic aerated lagoons.

Facultative Aerated Lagoons

Facultative aerated lagoons are those in which some solids may leave with the effluent stream and some settle down in the lagoon since aeration power input is just enough for oxygenation and not for keeping all solids in suspension. As the lower part of such lagoons may be anoxic or anaerobic while the upper layers are aerobic, these are termed as facultative aerated lagoons.

Further the facultative aerated lagoons are also known as partially mixed type aerated lagoons because these are operated at a low rate of aeration which is not adequate to keep all the solids in suspension.

Aerobic Aerated Lagoons

Aerobic aerated lagoons are those which are fully aerobic from top to bottom as the aeration power input is sufficiently high to keep all the solids in suspension besides meeting the oxygenation needs of the system. No settlement of solids occurs in these lagoons and under equilibrium conditions the new (microbial) solids produced in the system equal the solids leaving the system.

Thus in this case the solids concentration in the effluent is relatively high and some further treatment is generally provided after such lagoons. If the effluent is settled and the sludge recycled, the aerobic aerated lagoon, in fact, becomes an activated sludge or extended aeration type lagoon.

A few typical characteristics of the above types of aerated lagoons are given in Table.

Some Characteristics of Aerated Lagoons

S.No	Characteristics	Facultative aerated lagoons	Aerobic aerated lagoons	Extended aeration system (for comparison)
1.	Detention period, days	3–5	2–3	0.5–1.0
2.	Depth, m	2.5–5.0	2.5–4.0	2.5–4.0
3.	Land required, m²/person	0.15–0.30	0.10–0.20	—
4.	BOD removal efficiency, %	80–90	50–60	95–98
5.	Overall BOD removal rate constant, K at 20°C (soluble only), per day	0.6–0.8	1–1.5	20–30
6.	Suspended solids (SS) in unit, mg/l	40–150	150–350	3000–5000
7.	VSS/SS	0.6	0.8	0.6
8.	Desirable power level watts/m³ of lagoon volume	0.75	2.75–6.0	15–18
9.	Power requirement kWh/person/year	12–15	12–14	16–20

Facultative type aerated lagoons have been more commonly used the world over because of their simplicity in operation and minimum need of machinery. They are often referred to simply as 'aerated lagoons'.

Their original use came as a means of upgrading oxidation ponds overloaded due to industrial wastes without adding to the land requirement. Further as the aerated lagoons are deeper than the oxidation ponds, and as they are artificially aerated, less land and less detention period are required for aerated lagoons as compared to oxidation ponds.

Flow conditions in aerated lagoons are neither ideal complete-mixing nor ideal plug-flow in nature. They are dependent on lagoon geometry and are better described by dispersed flow models of the type given by Wehner and Wilhem for first order kinetics and hence the design procedure given below is based on this dispersed flow model which takes treatability of the waste, temperature and mixing conditions into account.

The aerobic aerated lagoons have a complete-mixing regime and a slightly different mode of design is followed. However, as aerobic aerated lagoons have not yet been built in India (except one case) further discussion is limited to facultative aerated lagoons only.

Suspension Mixed Lagoons

Suspension mixed lagoons flow through activated sludge systems where the effluent has the same composition as the mixed liquor in the lagoon. Typically the sludge will have a residence time or sludge age of 1 to 5 days. This means that the chemical oxygen demand (COD) removed is relatively little and the effluent is therefore unacceptable for discharge into receiving waters. The objective of the lagoon is therefore to act as a biologically assisted flocculator which converts the soluble biodegradable organics in the influent to a biomass which is able to settle as a sludge. Usually the effluent is then put in a second pond where the sludge can settle. The effluent can then be removed from the top with a low COD, while the sludge accumulates on the floor and undergoes anaerobic stabilisation.

Design of the Aerated Lagoons

Design Variables

For facultative aerated lagoons, the dispersed flow model gives the relation between influent and effluent substrate concentrations, S_0 and S, respectively and other variables such as the nature of the waste, the detention period and the mixing conditions, as shown in the Wehner-Wilhem equation given below-

$$\frac{S}{S_0} = \frac{4ae^{1/2d}}{(1+a)^2 e^{a/2d} - (1-a)^2 e^{-a/2d}}$$

in which the term $a = \sqrt{1 + 4Ktd}$

d = dispersion number (dimensionless)

$= (D/UL) = (Dt/L^2)$;

in which D = axial dispersion coefficient (length^2/time);

L = length of axial travel path;

t = theoretical detention period (volume/flow rate);

U = velocity of flow through lagoon (length/time);

K = substrate removal rate in lagoon (time^{-1}); and

S_0 and S = initial and final substrate concentrations (mass/volume)

A graphical solution of equation is shown in Figure from which it is seen that prior knowledge of the substrate removal rate K as well as of the mixing condition likely to prevail in a lagoon is necessary to determine the efficiency of BOD removal at selected detention period.

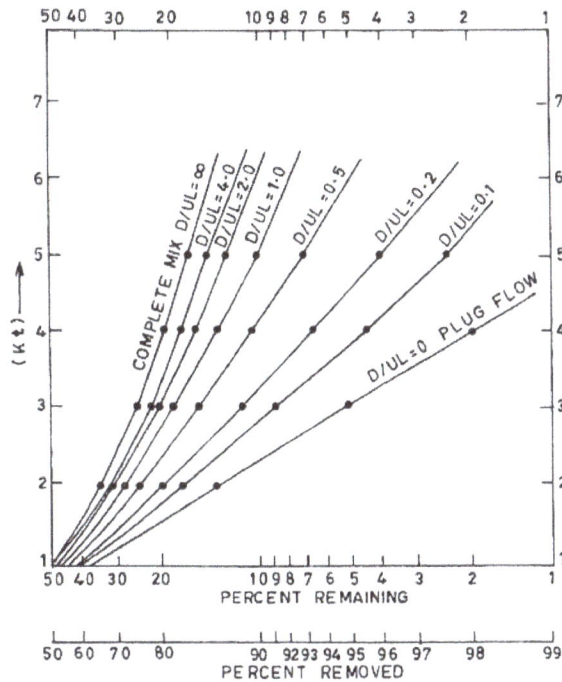

Graphical plot of Wehner-Wilhem equation for substrate removal
efficiency based on the dispersed flow model.

Mixing Conditions

The mixing conditions in a lagoon are reflected by the term d which is known as the dispersion number and equals (D/UL) or (Dt/L²). It is affected by various factors. Observed results have shown the (D/UL) values to be in the approximate range given in Table 15.3 for different length-width ratios of lagoons.

Likely Values of Dispersion Numbers (D/UL) at Different Length-Width Ratios

Aerated lagoon	Approximate range of (D/UL) values	Typical mixing condition
Length-width ratio 1:1 to 4:1	3.0 to 4.0 and over	Well mixed
Length-width ratio 8:1 or more	0.2 to 0.6	Approaching plug flow
Two or three units in series	0.2 to 0.6 (overall)	Approaching plug flow

By suitable choice of a lagoon's geometry one can promote either more plug flow or more complete mixing type of conditions. Figure shows some different types of arrangements using baffles or units in series. In case of units in series, each unit may be well mixed with value of D/UL approaching 3.0 or 4.0 but overall the arrangements would give a relatively plug flow type arrangement.

Values of D/UL can be determined by conducting dye (tracer) tests on existing units using well known methods, but where D/UL values are required for design purposes prior to construction; they can be estimated either from lab-scale models or by using empirical equations available.

Low values of D/UL signify plug flow conditions and generally give higher efficiencies of substrate removal whereas the converse is the case with higher values of D/UL. However, process efficiency is not the only consideration; process stability under fluctuating inflow quality and quantity conditions, has also to be kept in view.

For municipal or domestic sewage, relatively plug flow type conditions (i.e., low values of DU/L) are preferred. In case of industrial wastes, relatively well mixed conditions (i.e., higher values of D/UL) may be preferred depending upon the nature of industrial waste; the greater the fluctuations in quality and quantity of industrial wastes, the greater the advantage in adopting well mixed conditions.

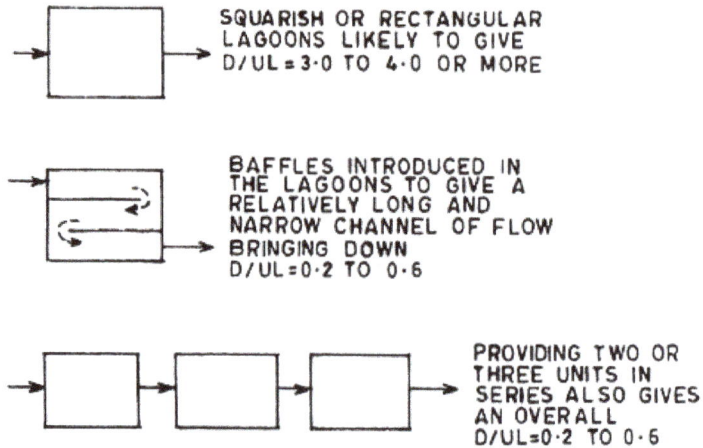

Estimated effect of lagoon geometry on value of dispersion number D/UL

Construction Details

Lagoons are generally rectangular in shape though it is not absolutely essential. Natural land contours may be followed to the extent possible to save on earthwork. Lagoon units may be built with different length-width ratios and arrangement of internal baffles to promote desired mixing conditions. Lagoons may also be provided as two or three stage systems with the subsequent units placed at a lower level than the first if desired.

Construction techniques for aerated lagoons are similar to those used in case of oxidation ponds with earthen embankments. Pitching of the embankment is desirable to protect it against erosion. In cases where soil percolation is expected, suitable lining may have to be provided to maintain the design level in the lagoon and avoid pollution of groundwater.

Substrate Removal Rates

As shown in Table for facultative aerated lagoons the overall substrate removal rate constant K for sewage at 20°C, i.e., K_{20} varies from 0.6 to 0.8 per day (soluble BOD basis). At any other temperature T°C in lagoon the value of K, i.e., K_T may be obtained from the following formula-

$$K_T = K_{20}(1.035)^{T-20}$$

The temperature in a lagoon T_L is estimated from the following equation

$$\frac{t}{h} = \frac{T_i - T_L}{f(T_L - T_a)}$$

in which

t = detention time (in days);

h = depth of lagoon (in m);

T_i = temperature of influent sewage (in °C);

T_a = temperature of ambient air (in °C); and

f = heat transfer coefficient (in m/day)

For aerated lagoons f = 0.49 m/day.

The average winter month temperature is critical for determining the detention time required. As stated earlier, the detention time to be provided in a lagoon can be determined from equation or Figure for any desired efficiency for the computed temperature and mixing conditions in the lagoon.

Power Level

The power input in facultative aerated lagoons has to be adequate only to diffuse dissolved oxygen uniformly in the system; no effort is made to keep the solids in suspension. Hence a minimum power level of 0.75 watts per m³ lagoon volume is adequate, but this should be checked with the aeration equipment supplier for its oxygenation characteristics and compatibility with proposed depth and shape of lagoon.

For treating domestic sewage the power requirement varies from 12 to 15 kWh per person per year or 2 to 2.5 hp per 1000 population equivalent. The oxygenation capacity of aerators is reported to range from 1.87 to 2.0 kg of oxygen per kWh at standard conditions for power delivered at shaft. Spacing of aerators should be adequate for uniform aeration all over the lagoon area without much overlap of the circle of influence of adjoining aerators as specified by the manufacturers. A minimum of two aerators would be desirable to provide to make up the total power requirement.

Aerators ranging from 3 to 75 hp are now readily available in our country. They can be either floating or fixed type. Floating aerators are mounted on pontoons (which should be corrosion-free). They have the advantage of being able to adjust themselves to actual levels obtained in the lagoons due to seepage and/or fluctuating inflows. Fixed aerators are mounted on structural columns and carefully levelled with regard to the outlet weir level to ensure required submergence of aerator blades to give the design oxygenation capacity.

Effluent Characteristics

The effluent is generally made to flow over an outlet weir. As the concentration of solids passing out in the effluent may be nearly the same as that in the lagoon the BOD corresponding to the volatile fraction of these solids (assumed as 0.77 mg per mg VSS in effluent) should be added to the value of the soluble BOD S obtained by use of equation or Figure.

Thus the final effluent BOD is given by-

Final BOD mg/l = S mg/l + (0.77) (VSS in effluent) mg/l

It is because of the suspended solids (expected to range from 40 to 60 mg/l in case of domestic sewage) in the final effluent that the total effluent BOD is difficult to reduce below 30 to 40 mg/1 in winter. At other times of the year BOD less than 30 mg/l may be possible.

This range of BOD is more than adequate for irrigation purposes, but for river disposal the applicable standards should be ascertained and design made accordingly. Where necessary, further reduction of BOD can be achieved either by a small increase in detention time or by more efficient interception of solids flowing out (e.g., providing deeper baffle plate ahead of outlet weir) or by provision of an additional treatment unit.

Nitrification is not likely to occur in aerated lagoons. Coliform removal ranges from 60 to 90% and it shows considerable seasonal variation.

Sludge Accumulation

Sludge accumulation occurs at the rate of 0.03 to 0.05 m³ per person per year as in the case of oxidation ponds and is manually removed once in 5 to 10 years and used as good agricultural soil. The depth of the lagoon may be increased a little to allow for sludge accumulation, if desired.

Methods of Aerating Lagoons or Basins

There are many methods for aerating a lagoon or basin:

- Motor-driven submerged or floating jet aerators;
- Motor-driven floating surface aerators;
- Motor-driven fixed-in-place surface aerators;
- Injection of compressed air through submerged diffusers.

Floating Surface Aerators

Ponds or basins using floating surface aerators achieve 80 to 90% removal of BOD with retention times of 1 to 10 days. The ponds or basins may range in depth from 1.5 to 5.0 metres.

In a surface-aerated system, the aerators provide two functions: they transfer air into the basins required by the biological oxidation reactions, and they provide the mixing required for dispersing the air and for contacting the reactants (that is, oxygen, wastewater and microbes). Typically, the floating high speed surface aerators are rated to deliver the amount of air equivalent to 1 to 1.2 kg O_2/kWh. However, they do not provide as good mixing as is normally achieved in activated sludge systems and therefore aerated basins do not achieve the same performance level as activated sludge units.

With low speed surface aerators SOTE (Standard Oxygen Transfer Efficiency) is higher thanks to better mixing capacity. This mixing capacity of an impeller depends highly on the impeller diameter. Low speed surface aerator present such high diameter. Therefore SOTE for low speed surface aerators is about 2 to 2.5 kg O_2/kWh. This is why low speed surface aerators are mostly used in sewage or industrial treatment as WWTP are bigger and sparing energy (and money) becomes very interesting.

Biological oxidation processes are sensitive to temperature and, between 0 °C and 40 °C, the rate of biological reactions increase with temperature. Most surface aerated vessels operate at between 4 °C and 32 °C.

Submerged Diffused Aeration

Submerged diffused air is essentially a form of a diffuser grid inside a lagoon. There are two main types of submerged diffused aeration systems for lagoon applications: floating lateral and submerged lateral. Both these systems utilize fine or medium bubble diffusers to provide aeration and mixing to the process water. The diffusers can be suspended slightly above the lagoon floor or may rest on the bottom. Flexible airline or weighted air hose supplies air to the diffuser unit from the air lateral (either floating or submerged).

Advantages of Aerated Lagoons

The various advantages of aerated lagoons are as indicated below:

(i)   The aerated lagoons are simple and rugged in operation, the only moving piece of equipment being the aerator.

(ii) The removal efficiencies in terms of power input are comparable to some of the other aerobic treatment methods.

(iii) Civil construction mainly entails earthwork, and land requirement is not excessive. Aerated lagoons require only 5 to 10 percent as much land as stabilization ponds.

(iv) The aerated lagoons are used frequently for the treatment of industrial wastes.

Decentralized Wastewater System

Decentralized wastewater treatment consists of a variety of approaches for collection, treatment, and dispersal/reuse of wastewater for individual dwellings, industrial or institutional facilities, clusters of homes or businesses, and entire communities. An evaluation of site-specific conditions is performed to determine the appropriate type of treatment system for each location. These systems are a part of permanent infrastructure and can be managed as stand-alone facilities or be integrated with centralized sewage treatment systems. They provide a range of treatment options from simple, passive treatment with soil dispersal, commonly referred to as septic or onsite systems, to more complex and mechanized approaches such as advanced treatment units that collect and treat waste from multiple buildings and discharge to either surface waters or the soil. They are typically installed at or near the point where the wastewater is generated. Systems that discharge to the surface (water or soil surfaces) require a National Pollutant Discharge Elimination System (NPDES) permit.

These systems can:

- Serve on a variety of scales including individual dwellings, businesses, or small communities;

- Treat wastewater to levels protective of public health and water quality;

- Comply with municipal and state regulatory codes; and

- Work well in rural, suburban and urban settings.

Need for Decentralized Wastewater Treatment

Decentralized wastewater treatment can be a smart alternative for communities considering new systems or modifying, replacing, or expanding existing wastewater treatment systems. For many communities, decentralized treatment can be:

- Cost-effective and economical

  o   Avoiding large capital costs

  o   Reducing operation and maintenance costs

  o   Promoting business and job opportunities

- Green and sustainable

  o   Benefiting water quality and availability

  o   Using energy and land wisely

  o   Responding to growth while preserving green space

- Safe in protecting the environment, public health, and water quality

  o   Protecting the community's health

  o   Reducing conventional pollutants, nutrients, and emerging contaminants

  o   Mitigating contamination and health risks associated with wastewater

The Bottom Line

Decentralized wastewater treatment can be a sensible solution for communities of any size and demographic. Like any other system, decentralized systems must be properly designed, maintained, and operated to provide optimum benefits. Where they are determined to be a good fit, decentralized systems help communities reach the triple bottom line of sustainability: good for the environment, good for the good people

Types

Based on the size of the served area, different scales of decentralization could be found:

- Decentralization at the level of a suburb or satellite township in an urban area – these systems could be defined as small centralized systems when applied to small towns or rural communities. But if they are applied only to selected suburbs or districts in medium or large population centres, with existing centralized system, the whole system could be defined as a *hybrid* system, where decentralization is applied to parts of the whole drained area.

- Decentralization at the level of a neighbourhood – this category includes clusters of homes, gated communities, small districts and areas, which are served by vacuum sewers.

- Decentralization at "on-site" level (on-site sanitation) – in these cases the whole system lays within one property and serves one or several buildings.

In locations with developed infrastructure, decentralized wastewater systems could be a viable alternative of the conventional centralized system, especially in cases of upgrading or retrofitting existing systems. Many different combinations and variations of *hybrid* systems are possible. The development of new treatment technologies allows for decentralized solutions, which are technically and aesthetically sound and acceptable.

Decentralized wastewater system in Torvetua eco-village in Norway.
Wastewater is collected by a vacuum sewer. Greywater is treated locally.

Decentralized applications are a necessity in cases of new urban developments, where the construction of the infrastructure is not ready or will be executed in future. In many countries and locations, the infrastructure development (roads, water supply and especially wastewater/drainage systems) is executed years after the housing development. In such cases decentralized wastewater facilities are considered as a temporary solution, but they are mandatory, in order to prevent public health and ecological problems.

Wastewater Treatment Options

Biogas digester for decentralized wastewater treatment at Meru Prison, Kenya

Treatment/disposal facilities requiring effluent infiltration: usually they are applied at on-site level and are adequate because of the very low wastewater quantity generated. However, they require suitable soil conditions, permitting infiltration of the excess water, and low ground water table. If not applied properly, they may be a serious source of ground water pollution.

- Pit latrines are applied when the water supply is very scarce and wastewater flow can hardly be generated. They are the most common sanitation technique in under-developed areas.

- Septic tanks are the most common on-site treatment technology used, which can be applied successfully where an adequate water supply is available and the soil/groundwater conditions are acceptable.

Treatment facilities resembling natural purification processes: their application requires significant surface area, because of the slow pace of the biological processes applied. For the same reason they are more suitable for warmer climates, because the rate of the purification process is temperature dependent. These technologies are more resilient to fluctuating loads and do not require complex maintenance and operation. Constructed wetlands are more suitable for applications at on-site or at neighbourhood level, while stabilization ponds could be a viable alternative for decentralized systems at the level of small towns or rural communities.

Engineered wastewater treatment technologies: there is a large variety of wastewater treatment plants where different treatment processes and technologies are applied. Small-scale treatment facilities in decentralized systems, apply similar technologies as medium or large plants. For on-site applications package plants are developed, which are compact and have different compartments for the different processes. However, the design and operation of small treatment plants, especially at neighbourhood or on-site level, present significant challenges to wastewater engineers, related to flow fluctuations, necessity of competent and specialized operation and maintenance, required to deal with a large number of small plants, and relatively high per capita cost.

In the specific case of developing countries, where localities with poor infrastructure are common, decentralized wastewater treatment has been promoted extensively because of the possibility to apply technologies with low operation and maintenance requirements. In addition, decentralized approaches require smaller scale investments, compared to centralized solutions.

Regulations and Management

Water pollution regulations in the form of legislation documents, guidelines or ordinances prescribe the necessary level of treatment, so that the treated effluent meets the requirements for safe disposal or reuse. Effluent may be disposed by discharging into a natural water body or infiltrated in the ground. In addition, regulations mention requirements regarding the design and operation of wastewater systems, as well as the penalties and other measures for their enforcement. Centralized systems are designed, built and operated in order to fulfil the existing regulations. Their management usually is executed by local authorities. In hybrid systems and small centralized systems in towns or rural communities management can be executed in the same way.

In the case of decentralization at on-site level and clusters of buildings, the whole wastewater system is located within private premises. The costs and responsibility for the design, construction, operation and maintenance is the responsibility of the owner. In many cases specialized companies might execute the operation and maintenance procedures. The local authorities issue permits and may provide support for the operation and management in the form of collecting wastes, issuing certificates/licenses for standardized treatment equipment, or for selected qualified private companies. From regulatory point of view, the control of the quality of treated effluent for reuse, discharge or disposal is entirely the responsibility of local or national government authorities. This might be a challenge if a large number of systems must be controlled and inspected. It is in the owner's interest to operate and maintain the system properly, especially in the case of reuse of the treated effluent. Most often the operational problems are associated with clogging of the treatment facilities as result of irregular removal of the sludge or hydraulic overloading due to increased number of population served or increased water consumption.

Urban Planning and Infrastructure Issues

Wastewater systems are part of the infrastructure of urban or rural communities and the urban planning process. Urban planning data and information, such as plots of individual dwellings, roads/streets, storm water drainage, water supply, and electricity systems are essential for the design and implementation of a sustainable wastewater system. In decentralized wastewater systems, which collect and treat wastewater only, storm water might be overlooked and cause flooding problems. If planned decentralized solutions are applied, storm water drainage should be executed together with the roads system.

In under-developed population centres where no infrastructure is available, is difficult to provide sustainable sanitation measures; e.g. pit latrines/septic tanks need periodic cleansing, usually executed by vacuum tracks, which have to access the latrine and need a basic road for this purpose. Fecal sludge management deals with the organization and implementation of this practice in a sustainable way, including collection, transport, treatment and disposal/reuse of faecal sludge from pit latrines and septic tanks.

In the cases of new urban/rural developments, or the retrofitting of existing ones, it is advisable to consider different alternatives regarding the design of the wastewater system, including decentralized solutions. A sustainable approach would require optimal technical solutions in terms of reliability and cost effectiveness. From this perspective, centralized solutions might be more appropriate in many cases, depending on existing sizes of plots, topography, geology, ground water tables and climatic conditions. But when applied adequately, decentralized systems allow for the application of more environmentally friendly solutions and reuse of the treated effluent, including recovery and reuse of specific elements. In this way alternative water resources are provided and the environment is protected. In addition public awareness, perceptions and support are an important part of the urban planning process in general, and the choice of adequate wastewater system in specific.

Examples

One example of decentralized treatment is the "DEWATS technology" which has been promoted under this name by the German NGO BORDA. It has been applied in many countries in South East Asia and in South Africa. It applies anaerobic treatment processes, including anaerobic baffled reactors (ABRs) and anaerobic filters, followed by aerobic treatment in ponds or in constructed wetlands. This technology was researched and tested in South Africa where it was shown that the treatment efficiency was lower than expected.

A case study of a decentralized wastewater system at on-site level with treated effluent reuse was performed at the Botswana Technology Centre in Gaborone, Botswana. It is an example of a decentralized wastewater system, which serves one institutional building, located in an area served by municipal sewerage. Wastewater from the building is treated in a plant consisting of: septic tank, followed by planted rock filter, bio-filter and a surface flow wetland. The treated effluent is reused for irrigation of the surrounding green areas, but the study registered outflow from the wetland only during periods of heavy rains. This example shows the need for careful estimation of the expected quantity, quality and fluctuations of the generated wastewater when designing decentralized wastewater systems.

Grinder Pump

A grinder pump is a waste management device. It is used to pump sanitary sewage from a building to the municipal sewerage system when the building's plumbing is at a lower grade than the sewer main or when there is not enough slope to allow sewage to gravity flow to the municipal system. Household waste from toilets, sinks and washing machines flows through the home's pipes into the grinder pump's holding tank. Once the waste inside the tank reaches a specific level, the pump will turn on, grind the waste into a fine slurry, and pump it to the municipal sewer system.

The components of a grinder pump system consist of the pump, a tank, and an alarm panel. The pump has a level sensing device called a "float". If the pump malfunctions and the waste level in the holding tank rises above a certain level, the alarm panel should alert the homeowner that the pump is experiencing problems. It is highly recommended that the alarm panel should have both an indicator light and an audible alarm.

Grinder pump systems are prone to grease buildup, which could lead to pump failure. Regular pump maintenance is highly recommended. Certain items such as sanitary napkins, diapers, kitty litter, paint, oil (both motor and cooking oils), should never be flushed or poured down any drain, whether the home is connected to a gravity sewer system, grinder pump system, or septic tank. Disposable wipes are not recommended for grinder pump systems.

Need for Grinder Pumps in some Homes/Businesses

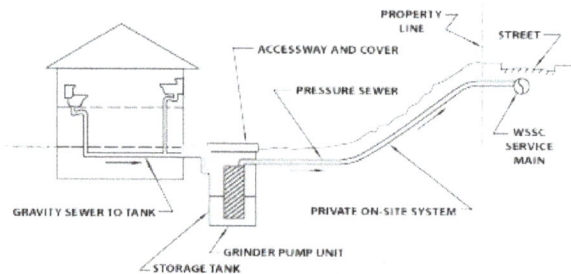

In most instances, wastewater flows by gravity from a home/business' on-property sewer service line to a public sewer main where it travels to wastewater treatment plants. At the plants, the wastewater is cleaned, disinfected and safely returned to the environment. However, because of elevation, gravity may not work in all instances. In situations where a home/business' sewer service line leaves the building at a lower elevation than the public sewer main, a grinder pump is sometimes used to grind and pump wastewater to the main.

Components

Grinder pump station with fiberglass tank and stainless steel lid installed outside a home

Grinder pump station with HDPE tank being installed

The grinder pump "station" consists of the pump, a tank, and an alarm panel. A pump for household use is usually 1 hp, 1.5 hp or 2 hp. A cutting mechanism macerates waste and grinds items that are not normally found in sewage, but may get flushed down the toilet. The pump has a level sensor either built into the pump, called "sensing bells," or attached externally to the pump, called "floats." (The level sensing devices vary among grinder pump manufacturers.) If the pump malfunctions and the waste level in the holding tank rises above a certain level, the alarm panel should alert the homeowner that the pump is experiencing problems. The alarm panel should have both a buzzer and an indicator light.

The holding tank, likely constructed of fiberglass, high-density polyethylene (HDPE) or fiberglass-reinforced polyester (FRP), has an inlet opening and a discharge opening. The pipes from the home are connected to the inlet; the pipe that leads to the sewer main is connected to the discharge. Often, more than one home or restroom (in a park, for example) can be connected to one grinder pump station. In this case, more than one inlet can be installed. It is a good idea to consult the manufacturer or factory representative before purchasing a grinder pump station to ensure that more than one inlet hole can be drilled.

The tank has a lid made from heavy-duty plastic or metal that is bolted and/or padlocked shut to prevent entry by unauthorized persons.

Maintenance

Grinder pumps should not require preventive maintenance. However, grinder pumps that use floats to sense the level in the holding tank are prone to grease buildup that may turn the pump on unnecessarily, or not turn on the pump at all, causing the tank to fill up and sewage to possibly back up into the home or yard. To prevent this, grinder pumps that use floats are often hosed down to remove the grease from the floats.

Homeowners are not usually limited by what they can or can not pour down their drains because their home has a grinder pump. Feminine hygiene products, diapers, kitty litter, paint, oil (both motor oil and cooking oils), etc. should not be flushed or poured down any drain, whether the home is connected to a gravity sewer system, septic tank, grinder pump or cesspool.

Disposable wipes that are made by cleaning companies for personal use, cleaning toilets, etc. are causing problems in communities around the United States. Not only do people clog their household plumbing, they are causing problems with household grinder pumps, lift stations, and sewage treatment plants. Some wipe companies say "flush one at a time," some say "not for pump systems," some say "safe for sewers." As recommended by Consumer Reports, wipes should be put into a garbage can instead of the toilet. The National Association of Clean Water Agencies has compiled a list of articles and municipal documents regarding wipes.

In large sewage pump stations, clogging problems are often avoided by installing a chopper pump in the tank. A chopper pump is able to handle larger/tougher solids than a residential grinder pump, including hair balls, diapers, sanitary napkins, clothing, etc.

Grease Trap

Grease traps have been in use since Victorian times, with the first patent for a modern day grease trap issued to Nathaniel Whiting in the late 1800s.

Grease traps vary in size, with smaller variations designed to connect individual sinks and larger ones installed to service larger facilities.

The definition of a grease trap is "a trap in a drain or waste pipe to prevent grease from passing into a sewer system."

A grease trap is in simple terms a plumbing fixture that contains decomposing food waste.

Types of Grease Interceptors

There are four major types of grease interceptors found in most food service establishments:

- Small passive Hydromechanical Grease Interceptors (HGI)

Hydromechanical grease interceptors (most often referred to as grease traps) often sit inside the kitchen underneath the sink or in the floor. They passively trap grease over time and need to be pumped frequently, as often as every week for high producing sites. While they can be less expensive up front, they do cost more to clean due to the frequency.

Most HGI's must be certified to a standard such as the ASME A112.14.3 or PDI-G 101 which require interceptors to retain a certain percentage of the grease which passes into the tank.

- Gravity Grease Interceptors (GGI)

Gravity Grease Interceptors are often made out of concrete, but can be made out of steel, fiberglass, and plastic. These interceptors are often greater than 500 gallons in liquid capacity, but hold a small percentage of their liquid capacity in grease and are not certified to meet any efficiency standards. This is why cities require they be pumped out once the grease and solids amount to 25% of the contents. Gravity Grease Interceptors are usually pumped every 90 days and can cost hundreds of dollars in maintenance each year. GGI's made of concrete can fail as quickly as every 15 years depending upon the quality of the materials.

- Automatic grease/oil recovery systems such as Big Dipper

Automatic grease removal devices or recovery units offer an alternative to Hydromechanical Grease Interceptors in kitchens. While their tanks passively intercept grease, they have an automatic or active mechanism for removing the grease from the tank and isolating it in a container. These interceptors must meet the same efficiency standards as a passive HGI, but must also meet a standard such as the ASME A112.14.4 which proves they are capable of skimming the grease effectively.

The upfront cost of these units is higher, but maintenance can be handled by the kitchen staff, eliminiating regular pumping charges and saving on operational costs.

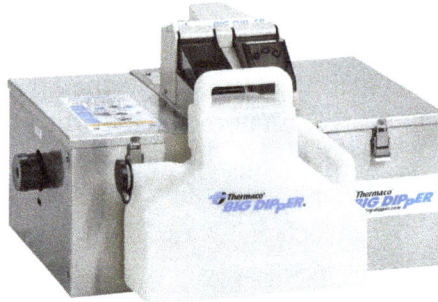

- Maximum Retention Hydromechanical Grease Interceptors such as Trapzilla

Max retention HGI's have become more popular in recent years as restaurants are opened in non-traditional sites without the space for a Gravity Grease Interceptor. These interceptors take up less space and hold significantly more grease per their volume than other alternatives, often between 70-85% of their liquid capacity. Most are made out of plastic while some are manufactured out of fiberglass.

Like other hydromechanicals, these interceptors must meet efficiency standards and most manufacturers test beyond the minimum standard to the fail point to demonstrate the full capacity of the unit.

Traditional passive grease trap designs date back to 1885 when the first U.S. patent was issued. Today's large and small grease interceptors use the same basic operating design as the 1885 model. While they do capture some grease, they are often inefficient at retaining grease and removing the grease is a task left to the owner.

Small passive traps must be cleaned out by hand. Large pre-cast traps, on the other hand, must be cleaned out by a professionally operated vacuum or pump truck.

Need for a Grease Trap

While sewer collection systems exist to take waste water to a treatment plant, there are some things is just was not designed to handle. One of those is grease.

Grease, especially grease with animal fats, cools and solidifies at normal temperatures in pipes. When this happens, blockages can form in the sewer pipes, eventually causing backups in the collection system called sanitary sewer overflows (SSO's). SSO's are a significant health risk to the public, so it's in everyone's best interest to keep our pipes clear of grease.

For this reason, many cities require the use of grease traps, more technically referred to as grease interceptors at locations that prepare food items to ensure grease does not ultimately cause sanitary sewer overflows.

In some instances, installing a grease interceptor can save the food service establish as well. Any location with long plumbing runs to the sewer collections system, such as a mall, hospital, or restaurant inside of a large building is in danger of creating blockages in the internal pipes which could lead to backups, fines, and perhaps even downtime as the internal plumbing is repaired.

Even If a restaurant has a traditionally designed passive trap which is not cleaned out on a timely basis, it will begin to allow too much grease into the sewer system, where it can cause blockages and sewage backups. This creates problems for wastewater system operators as well as the food service establishment.

Improperly maintained grease traps (or the failure to install a grease trap when required) often leads to fines, down-time, and can lead to bad publicity.

Working of a Grease Interceptor

In the most fundamental term, it functions by cooling warm or hot greasy water. By allowing the FOG to cool, it's able to separate the different layers of wastes.

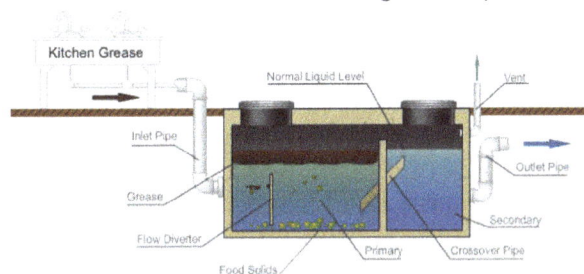

How a grease trap works

FOG are lighter in density and float at the surface of the tanks, and the cooled water – minus the FOG – continues draining to the sewer.

The FOG floating at the surface of the water is then trapped by the grease trap. For its use to be effective, the grease trap should be emptied when the sludge level reaches 25% of the tank level.

The larger "gravity" or "passive" traps rely more on gravity and time to separate FOG. Stand-pipes, internal baffling, and larger tank sizes allow the magic of gravity to work by increasing the "retention time" on these passive units. The services of a professional should often be sought at intervals.

For smaller, indoor types a hydro-mechanical grease interceptor should be used instead. The traps used in the interceptor utilize internal baffles that provide more space and time for separation by lengthening the flow path.

In order to maximize separation of FOG from effluents, the wastewater making its way inside the grease trap should be regulated by control devices that at this stage also allow mixing of air.

Grease Trap or Interceptor Sizing

Size is, needless to say, the most important thing for an establishment to consider. This is tied to the rating that hydro mechanical interceptors have in terms of the allowable maximum drainage in gallons per minute (GPM). Naturally, hydro mechanical receptors have a 100gpm handling capacity, and anything beyond this should be left to the gravity receptor.

Things that influence the grease trap/interceptor sizing are:

- Sizing as per Values of Drain Fixture Unit (DFU)
- Sizing as per the volume of total flowing fixtures
- Sizing based on waste pipe's diameter

To this end, it becomes necessary to have a manufacturer that provides charts with GPM flow listing based on the diameter of the pipe.

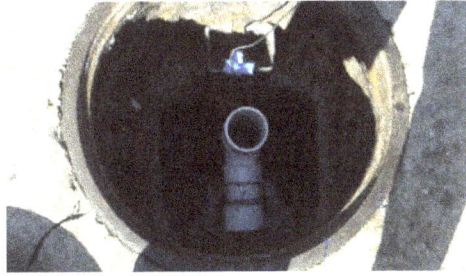

Provided you know the waste pipe's size and type of interceptor, this method becomes very simple. Similarly, using DFU values for sizing is straightforward because one only has to sum up the DFU values for the waste flowing into the grease trap.

Some basic math is required when calculating the interceptor's size based on the flow and capacity of the actual waste.

First, you'll need to determine the fixture's dimensions, which will, in turn, determine the fixture's volume in cubic inches.

Secondly, that figure would need to be converted to gallons. To better illustrate this, divide the volume aforementioned (cubic inches) by 231 to find the number of gallons of waste.

Thirdly, multiply the figure resulting from above by 75%. This is because it is very unlikely that the fixture will fill to the brim even on busy occasions.

Finally, calculate the flow/drain rate.

So how is a flow rate calculated?

You'd need to fill the fixture to 75% capacity, which is 3/4 of the total fixture. Next, time the length of period it takes for the fixture to drain completely.

Lastly, take the result you had in the third step above and divide it by the time it takes for the fixture to come out. The result you get is now the rate in GPM.

At this stage, it would be advisable to seek the guidance of a professional or a manufacturer. Also, depending on the local laws, you may need to consult the local health and hygiene department to be sure that the grease trap you have fits your needs.

Too large, and you run the risk of damaging things downstream as sulfuric acid may be created within the tank.

Too small, and the unit will not do its purpose of preventing FOG from passing through freely unless it's periodically cleaned.

Furthermore, the Code of Uniform Plumbing that acts as a guideline in this space prescribes that unless specifically required by the jurisdiction's authority, dishwashers are not to be attached to a grease trap.

This is because chemicals and other detergents used can emulsify the FOG rendering the trap unusable.

Again, the Code only permits garbage disposals to flow unhampered into drainage systems. It goes without saying that it is imperative to check with the local authority before purchasing or getting down with the sizing process.

Grease Trap Efficiency

While traditional grease traps and gravity interceptors have been the standard for more than 100 years, recent updates in this field are having a critical impact on the design and implementation of separator technology. The size of concrete, steel, or fiberglass gravity grease interceptors makes them difficult and costly to locate, especially in urban environments. Additionally, their retention efficiencies make them less effective as separators while newer technology allows for greater storage and less stagnant water which can easily turn into hydrogen sulfide.

As grease traps of the traditional design fill with grease, their efficiency at separating grease from wastewater decreases. When a trap is filled to capacity with fats and oil, separation no longer occurs and the trap no longer functions properly.

Traditional concrete traps must be pumped out after just 25 percent of their volume is filled, because after that they no longer work well enough to keep fats and oils out of the sewage systems.

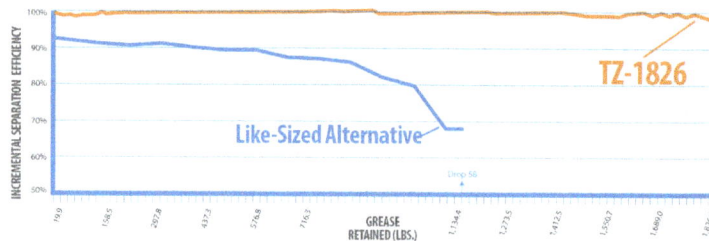

Best in Class Separation & Retention Efficiency

Automatic Grease Interceptor Efficiency

Big Dipper systems work a bit differently than traditional traps. In them, grease is skimmed out automatically on a programmed schedule based on the amount of grease produced.

This automation means employees don't have to measure or check grease levels. The grease in these automatic systems accumulates in a separate chamber and is simply disposed of in a municipally approved waste container.

While more expensive to purchase, automatic grease interceptors give Food Service Establishments control over their grease waste management and save customers from paying service companies on a weekly or monthly basis to clean out the grease interceptor.

Things that can go Wrong with Grease Traps

If the trapped FOG is not pumped out when needed, it becomes thick and may begin to escape via the outlet tee.

Also, when blocking occurs in the downstream pipes, the grease traps and drains containing the wastewater could flood out.

This could consequently lead to your kitchen being shut down by local sanitary officials.

How Often Should you Pump your Grease Interceptor?

For correct functioning, a deep layer of water is needed. To this end, it's recommended that a grease trap be drained as soon as the food layers and grease occupy 25% of the trap's space.

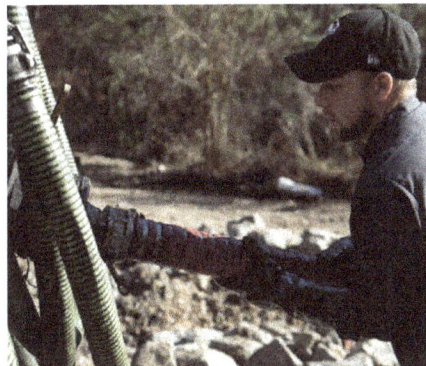

Pumping is Dependent on

Cleaning Practices

To reduce the amount of FOG going down the drain, it's necessary that dishes be scraped before being washed. As a result, less food and grease will make its way down the drain. Again, as a standard practice, high-fat liquids and fryer oil should never be allowed to flow down the drains.

The Type of Food

The more grease in the waste, the more often it needs to be pumped. Waste in the form of frostings, sauces, meats, salad dressing, and butter oil means a higher frequency.

The Amount of Food

Busy establishments lead to more waste products being washed down the drains. For this reason, busy seasons mean a higher pumping frequency. Needless to say, busy season isn't a convenient time for having your kitchen flooded.

The Trap or Interceptor Sizing

It goes without saying that the smaller the trap, the more often it will need to be pumped.

References

- Tchobanoglous, G.; Burton, F.L.; Stensel, H.D. (2003). Wastewater Engineering (Treatment Disposal Reuse) / Metcalf & Eddy, Inc (4th ed.). McGraw-Hill Book Company. ISBN 0-07-041878-0

- Ashworth, J; Skinner, M (19 December 2011). "Waste Stabilisation Pond Design Manual" (PDF). Power and Water Corporation. Retrieved 11 February 2017

- Anderson K., Dickin S., Rosemarin A. (2017) Towards "sustainable" sanitation: challenges and opportunities in urban areas", Sustainability, 8, doi: 10-3390/su8121289

- Karman D. (2007) The 'Cloaka Maxima' and the monumental manipulation of water in archaic Rome, On-line journal The Water of Rome, retrieved on the 18 March 2017

- Middlebrooks, E.J. (1982). Wastewater Stabilization Lagoon Design, Performance and Upgrading. McMillan Publishing. ISBN 0-02-949500-8

- Ashworth, J; Skinner, M (19 December 2011). "Waste Stabilisation Pond Design Manual" (PDF). Power and Water Corporation. Retrieved 11 February 2017

Permissions

We would like to thank the editorial team for lending their expertise to make the book truly unique. They have played a crucial role in the development of this book. Without their invaluable contributions this book wouldn't have been possible. They have made vital efforts to compile up to date information on the varied aspects of this subject to make this book a valuable addition to the collection of many professionals and students.

This book was conceptualized with the vision of imparting up-to-date and integrated information in this field. To ensure the same, a matchless editorial board was set up. Every individual on the board went through rigorous rounds of assessment to prove their worth. After which they invested a large part of their time researching and compiling the most relevant data for our readers.

The editorial board has been involved in producing this book since its inception. They have spent rigorous hours researching and exploring the diverse topics which have resulted in the successful publishing of this book. They have passed on their knowledge of decades through this book. To expedite this challenging task, the publisher supported the team at every step. A small team of assistant editors was also appointed to further simplify the editing procedure and attain best results for the readers.

Apart from the editorial board, the designing team has also invested a significant amount of their time in understanding the subject and creating the most relevant covers. They scrutinized every image to scout for the most suitable representation of the subject and create an appropriate cover for the book.

The publishing team has been an ardent support to the editorial, designing and production team. Their endless efforts to recruit the best for this project, has resulted in the accomplishment of this book. They are a veteran in the field of academics and their pool of knowledge is as vast as their experience in printing. Their expertise and guidance has proved useful at every step. Their uncompromising quality standards have made this book an exceptional effort. Their encouragement from time to time has been an inspiration for everyone.

The publisher and the editorial board hope that this book will prove to be a valuable piece of knowledge for students, practitioners and scholars across the globe.

Index

www.ingramcontent.com/pod-product-compliance
Lightning Source LLC
Chambersburg PA
CBHW082057190326
41458CB00010B/3519